The Worm Book

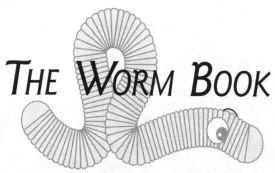

THE WORM BOOK

The Complete Guide
to Worms in Your Garden

LOREN NANCARROW
AND JANET HOGAN TAYLOR

TEN SPEED PRESS
Berkeley

All rights reserved. Published in the United States by Ten Speed Press, an imprint of
the Crown Publishing Group, a division of Random House, Inc., New York.
www.crownpublishing.com
www.tenspeed.com

Ten Speed Press and the Ten Speed Press colophon are registered trademarks of
Random House, Inc.

Library of Congress Cataloging-in-Publication Data
Nancarrow, Loren
The worm book: the complete guide to worms in your garden
/ by Loren Nancarrow and Janet Hogan Taylor
          p.   cm.
          Includes index.
          1. Earthworms. 2. Earthworm culture.
3. Gardening. I. Taylor, Janet Hogan, 1954- . II. Title.
SB998-E4N35 1998 97-48841
639'.75—dc21   CIP

ISBN-13: 978-0-89815-994-3 (pbk.)

Text design by Lisa Patrizio
Cover design by Gary Bernal
Illustrations by Janet Hogan Taylor

First Edition

146028962

Acknowledgments

The authors wish to acknowledge their gratitude to the following people (and others): Julie Castiglia, our agent and friend, who believed in us and made our books possible; Don Trotter (alias Dr. Curly), for his love of things that grow and his willingness to share his great knowledge when we really need it; our families—Brian, Evan, Leah, and Sue, and Susie, Graham, Hannah, and Britta—who have learned more about earthworms than any two families should have to know; the television viewers of San Diego, who have eagerly field-tested our data and ideas; and finally, to earthworms everywhere, who surely do work harder than us.

Thank you all!

CONTENTS

1
WHY DO WE NEED WORMS?

On the small farm we tend in San Diego, California, earthworms do most of the work. The farm has a 3,000-square-foot vegetable garden. We don't use rototillers because they compact the soil. Instead, earthworms do the tilling. They also help fertilize our crops, condition the soil, and eat our leftovers. In our orchard, they keep the soil near the fruit trees loose and rich. Earthworms are at work twenty-four hours a day helping to keep the farm a showplace.

Beyond the horticultural advantages, earthworms provide a diversion for children in the garden. Kids will happily dig and collect earthworms long enough to get some weeding done.

Simply put, earthworms are the hardest working creatures on (or under) the earth. They are worthy of our respect and admiration, and yet historically they've evoked fear and loathing. After all, it's worms that crawl in and out and eat our snout, and worms that we have to eat if nobody likes us. It's a shame that a couple of unfortunate children's songs lead us to think poorly of such magnificent organisms. These remarkable workers have many important roles in nature, including mixing and aerating the soil, improving soil structure and water infiltration, helping moderate soil pH, bringing up minerals in the soil, making nutrients more available to plants, breaking down plant and animal material into compost, and increasing beneficial microbial action in the soil. These are no small tasks, but the earthworm accomplishes them easily through its daily feeding habits.

Do you want to get rich? There are many people who will tell you that earthworms can help you do it beyond your wildest dreams. Don't believe it! Do believe this: there are not enough earthworms in the soil today. Regular plowing and spraying, disturbing the soil, and soaking it with chemical fertilizers and pesticides all take their toll on earthworms. Society is beginning to learn what damage these substances do to the soil and its inhabitants; now we can begin the long task of rejuvenating the soil. Perhaps the best thing you can do to help is to grow some earthworms in a garden bed outside or in a bin under your kitchen sink. Sell some if you'd like, or sell their castings instead, or just grow them to return to the soil, which is so in need of your earthworms' labor. Along the way you'll learn a bit of husbandry and biology. You'll be amazed at an earthworm's ability to convert what we think of as garbage into gold, and you'll be doing your part to put the natural order back in place. Read on to learn how earthworms work their magic—and how you can be a part of it.

THE SOIL

There are many different types of soil all around the world. Soils can be loamy, sandy, or clay/adobe, just to name a few, but soil itself is made up of two main parts. One part is made up of rock particles that at one time or another were part of a larger rock or stone. Over time, erosion of rocks and stones by wind and water produces soil particles. (An example of this kind of soil particle is sand. If you look closely at sand, each particle looks like—and is—a miniature rock.)

The other part of soil is decaying organic material. As plants and animals die and decompose, they are broken up into smaller particles called humus. It's the humus part of soil that holds water, feeds plants, and keeps the soil from becoming too hard for plants to grow in. By eating and breaking down large pieces of decaying matter, earthworms play a key role in increasing the humus in soil.

The United States Department of Agriculture decided to test fertilizer versus earthworms over forty years ago. To do this, the department started with two containers of poor soil. To one container they added dead worms, fertilizer, and grass seed. To the other container they added live worms and grass seed—no fertilizer. To their amazement, the grass seed in the container with the live worms grew four times faster than the grass seed in the container with dead worms and fertilizer.

It is estimated that in an area with large numbers of earthworms, the worms can cover an acre of land with as much as eighteen tons of new soil each year; but it is also estimated that we are using seventeen times more topsoil than is being produced.

Earthworms are essential in good soil composition. As they burrow through the soil, they open it up and help keep it loose. This tilling action allows oxygen and water to get down into the soil where they can be taken up by plants; these elements in turn improve soil conditions for beneficial bacteria and other microorganisms that contribute to healthy soils. Earthworms also bring up soil from deeper soil levels to the top and then bring topsoil back down again. Over time, soil that is brought up by worms will cover seeds and allow them to germinate. This process can bury rocks and other objects.

Plant roots have an easier time getting down into the soil when they follow earthworm burrows. Nitrogen-fixing bacteria, needed by plants for growth and vigor, have been found in large numbers along the sides of earthworm burrows.

When earthworms feed, they take in bits of rock and organic matter (humus), digest what they can, and deposit the rest as excrement (castings). Earthworm castings improve the soil in several ways:

- Castings are close to neutral in pH—around 7 on the pH scale—no matter what kind of soil the worm ate. For example, even if a worm fed in a very acidic soil, its castings would be neutral, not acidic. Earthworm castings also contribute to neutralizing soil pH by adding calcium carbonate to the soil.

- Castings are rich in minerals and nutrients needed by plants. A study at Cornell University showed that the

> A nightcrawler is very strong for its size. A nightcrawler that weighs only $1/13$ of an ounce, has been shown to move a stone that weighs 2 ounces. That is equivalent to a 200-pound man moving over $2\frac{1}{2}$ tons.

nutrient level of castings is usually much higher than that of the surrounding soil. Castings were found to be high in nitrogen, potassium, phosphorus, magnesium, and trace minerals. Castings were also shown to supply needed micronutrients to plants. Another study estimated that castings contain five times the available nitrogen, seven times the available potash, and one and a half times the calcium found in good topsoil. So castings are excellent plant fertilizers and provide nutrients in a form immediately available for plant use.

- Castings are food for other beneficial microorganisms. They will contain thousands of bacteria, enzymes, and remnants of plant and animal material that were not digested by the earthworm. The composting process then continues long after the casting is excreted, adding beneficial microorganisms back to the soil and providing a source of food for the ones already there. Some of these soil organisms release potassium, phosphorus, calcium, magnesium, iron, and sulfur into the soil ready for plant use.

- Castings increase the humus content of the soil. An excreted casting is 65 to 70 percent organic matter, or humus. Soil rich in humus soaks up and holds water better. The soil is loose and is less likely to become hard and compacted. Humus can also buffer soil by binding with and holding the heavy metals from materials such as manure, sewage sludge, and vegetable waste matter (stems and roots) left over from crops.

- Castings hold their nutrients in mucus membranes that are secreted by the earthworm. This allows the nutrients to be slowly released so they are available to the plants over a period of time as needed.

C:N RATIO AND WORMS

Plants must have a way to take in the minerals they need from their soil environment. Scientists have discovered that for this assimilation to occur, a certain ratio of carbon to nitrogen (C:N) must exist. Looking at fallen leaves provides an interesting example. Several studies have measured the carbon to nitrogen ratio of many common tree species, and in no case does a tree's leaf litter come close to the optimum 20:1 ratio needed by plants. Most trees have too high a carbon content. A few examples are: 24.9:1 for elms, 42:1 for oaks, and a whopping 90.6:1 for Scotch pines. So how can the dead leaves be converted into decomposed organic matter that has the correct ratio for plants to use?

When plant litter breaks down and decomposition has started, nitrogen and carbon levels decrease with each decomposer that feeds on it. Carbon is a food source and therefore decreases more quickly than nitrogen.

Earthworms play a big role in this breakdown. When an earthworm feeds on leaf litter and breaks the litter down during metabolism, the carbon level falls. The earthworm castings may still have a C:N ratio too high for plants to directly use the nitrogen, but then other decomposing organisms can use the castings for food. The castings are further broken down, and, when the resulting organic matter has a 20:1 ratio, plants will be able to directly use the nitrogen the leaves contained.

RECYCLING

Earthworms are excellent composters. They can compost organic material faster than any composting system. Some earthworm species will eat half their body weight in food per day. The nightcrawler will come out at night and search for plant matter it can pull back into its burrow. Once the food is pulled in and eaten, the nightcrawler will deposit its castings back on the surface of the soil. The castings in turn become fertilizer for plants. So, for example, if you mow your lawn with a mulching mower—one that returns the clippings to the lawn—earthworms can find and eat the clippings and spread their castings through the top of the soil. This is a simple example of recycling the clippings' nutrients back to the lawn—but the benefits of recycling with earthworms don't stop there.

Earthworms can be maintained in a controlled situation to compost household, yard, and animal wastes. The homeowner can easily maintain a household worm bin to take care of kitchen wastes. A gardener can use earthworms directly in his garden soil or in an outdoor worm bin to help compost plant material. Finally, animal wastes can also be composted into rich vermicompost that can be used on garden plants. Approximately 70 percent of the material we send to landfills, including kitchen wastes, farmyard manures, and yard wastes, can be used to feed worms. If we did feed this material to the worms, the worms could give us 60 percent of the volume back as vermicompost fertilizer. This fertilizer would be a safe, natural soil enhancer and plant food that would be a benefit to the environment.

2

WHAT IS A WORM? BASIC WORM BIOLOGY

When describing an earthworm to someone who has never seen one, it sounds like you are describing a creature that is too good to be true and can't possibly exist. They don't have any ears, eyes, or a nose, but they do have senses. They have a mouth, but they don't have jaws or teeth. Each earthworm is both male and female—but it still takes two earthworms to make little earthworms. Earthworms are truly specialized creatures, perfectly adapted to subterranean life, and they excel at turning the stuff we would consider waste into a useful product.

EARTHWORM HISTORY

Charles Darwin, father of evolutionary theory, said of the earthworm, "It may be doubted whether there are many other animals in the world which have played so important a part in the history of the world." Darwin was fascinated by earthworms and studied them for thirty-nine years. He even wrote a book about earthworms, called *The Formation of Vegetable Mould Through the Action of Worms With Observations on Their Habits.*

Earthworms are members of the phylum Annelida, or segmented worms. This phylum has three classes, with earthworms belonging to the class Oligochaeta, of which there are around six thousand known species. It is thought that earthworms arose during the Cretaceous era, when dicotyledonous plants appeared, but some evidence suggests they arose in the much earlier Jurassic period. Most scientists agree that earthworms have been on Earth for at least 120 million years.

Earthworms have been well recorded in history, and not just by Darwin. The Greek philosopher Aristotle called earthworms "the intestines of the soil." He wasn't far off with this observation. Even in the time of Egyptian pharaohs, Cleopatra herself said, "earthworms are sacred." With a history like this, why don't earthworms get more respect?

In North America earthworms have had their ups and downs. Scientists believe that most of the earthworm species were killed here in the last ice age, about ten to fifty million years ago, by glaciers that dipped down from the Arctic into the temperate regions. But, you may be thinking, you have seen earthworms in your very own yards. That's because earthworms were reintroduced to North America by early European settlers in the seventeenth and eighteenth centuries. Most worms arrived in the soil clinging to the roots of favorite plants brought to settle the new land. The settler's ships also used soil as ballast, and this was off-loaded at ports once it was no longer needed. The soil contained many earthworms, which gradually spread out from the many ports. Some farmers, after seeing plants in the port cities do better with the earthworms, deliberately introduced the earthworms to their land.

In many localities throughout the world, and in particular the southern hemisphere, man has played an important part in the introduction of earthworm species. A study of earthworm species in several large cities in Chile found that all the earthworm species there originated from Europe. Of the nineteen earthworm species presently found in Canada, only two of them are thought to be indigenous. The rest are imports.

The endemic *Lumbricus,* the genus of nightcrawlers and some redworms, have been found to form a belt around the temperate regions of Europe, Asia, and eastern North America.

FIVE HEARTS AND NO LEGS:
THE BODY STRUCTURE OF AN EARTHWORM

Earthworms are cold-blooded invertebrates and hence have no backbones. Instead their bodies are broken down into segments that vary in width, with the largest being in the front region of the worm. The segments are numbered and scientists use the numbers to differentiate among earthworm species.

Mature worms have a structure called a clitellum. This structure is the glandular portion of the epidermis, or skin, which is associated with cocoon formation. The clitellum can differ widely among different species. Sometimes it appears as a swollen area, and in others as a well-defined constriction in the worm. The clitellum can be a different color than the rest of the worm—usually darker or lighter in tone, but sometimes a completely different color. The position of the clitellum on the body of the worm differs in each species as well. In *Lumbricus*, the clitellum is positioned between segments twenty-six and thirty-two on the anterior or top part of the body.

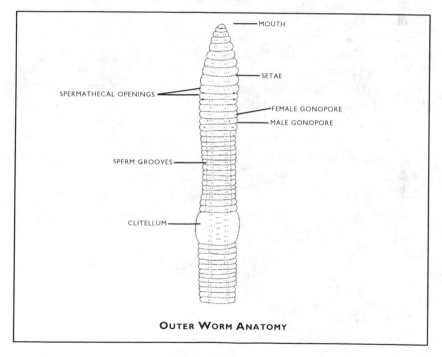

OUTER WORM ANATOMY

On every segment except the first segment, earthworms have bristles (setae). There are four pairs of setae per segment for the earthworm, *Lumbricus*, but this number varies with species. The setae, which appear in a variety of shapes and lengths, come from exterior follicles on the body wall. Most of the setae on *Lumbricus* are curved in shape and approximately one millimeter in length.

The primary function of setae is locomotion, but they also play a role in reproduction.

To move, the earthworm extends its body, anchors it with its setae, and then contracts its body using its longitudinal muscle. Each extension, anchorage, and contraction is called a step. During this process, each segment can move forward two to three centimeters; the worm can take seven to ten steps per minute. There are several different kinds of pores located on a worm's body. Usually earthworms have two kinds of pores for reproduction: spermathecal and female. In addition, worms have dorsal pores, which are small openings in the segmental grooves of the worm. These pores are excretory structures for secreting coelemic fluid (what we know as worm slime). Some worm species have a defense mechanism where, when the worm is threatened, it can shoot a stream of mucus several centimeters in the air! I know that would get my attention!

Earthworms' bodies consist of 75 to 90 percent water, but are high in protein, making them a favorite food of moles, shrews, and birds.

Finally, small nephridiopores located on the ventrolateral surfaces of each segment are the openings of the nephridia (the excretory organs of the worm); these remove liquid wastes from the body.

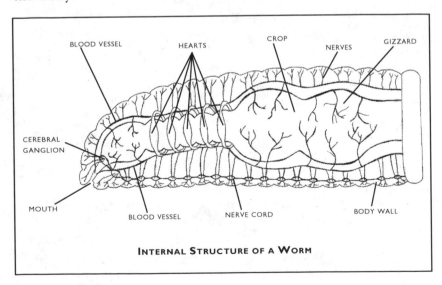

INTERNAL STRUCTURE OF A WORM

The body wall itself consists of an outer cuticle called the epidermis, which is very thin and helps to prevent water loss. In this layer the mucus or goblet cells can be found. They secrete the mucus that covers the body of the worm. Underneath the epidermis is a layer of nervous tissue containing large numbers of sensory cells that respond to stimuli such as touch, heat, and light. The epidermis and the nervous tissue are bound together by a basal membrane. Inside the membrane there are two muscle layers: One is a circular layer that goes around the worm's body and the other is a longitudinal muscle layer that is thicker and runs the length of the worm's body. Finally, the peritoneum, a layer of coelomic epithelial cells, separates the body wall from the body cavity.

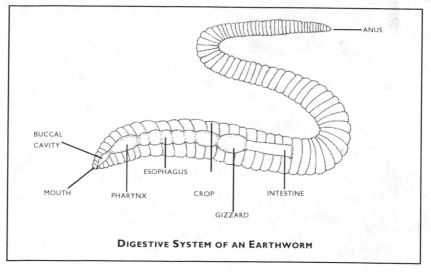

DIGESTIVE SYSTEM OF AN EARTHWORM

Worms don't have a defined head, but we consider the end with the mouth to be the head and the end with the anus to be the tail. We call the head the anterior and the tail the posterior.

Food—consisting of bits of organic matter mixed with soil— is taken in as the worm moves in the soil. Many worms prefer to feed where soils are rich with dead plant roots, dead leaves, decomposing plant matter, animal feces, or soil microorganisms. The food is picked up by the mouth, a small fleshy pad called a prostomium contracts over the mouth, and the food gets pulled into the alimentary canal. This canal is nothing more than a tube that extends from the mouth to the anus. Along the way, the food passes different sections of the tube, which help to break the food down. These sections are the buccal cavity, pharynx, esophagus, crop, gizzard, and intestine.

So does an earthworm "hear," "see," or "smell"? Yes and no. Like a snake, the earthworm uses its setae to sense vibrations and "hear." The body wall contains many nerve receptors that taste chemical changes (or "smell") and other nerve receptors that detect light changes (or "see") in their environment. One interesting fact is, earthworms can't "see" the color red.

The buccal cavity is a small cavity (like the inside of an animal's mouth between the mouth opening and the pharynx) that has neither jaws nor teeth. The pharynx is thick and muscular and acts as a suction pump, drawing in food and pushing it down the canal. The esophagus starts out as a tube leading from the pharynx and becomes the crop and gizzard. The crop and gizzard may sound familiar to you because both of these are also found in birds. These structures basically have the same function in the earthworm as they do in the birds. The crop stores food and the gizzard grinds the food up. The rest of the alimentary canal is the intestine, where digestion and absorption of food nutrients take place. Finally, food and soil that are not digested are excreted through the anus as a worm manure called castings.

Lying alongside the intestine are narrow blood vessels that absorb the nutrients from the alimentary canal and feed the rest of the body. They extend almost the entire length of the worm's body. Between the blood vessels in the upper quadrant of the worm's body can be found anterior loops of vessels. These vessels ("hearts") are enlarged, have the ability to contract, and contain valves. *Lumbricus* has five pairs of such "hearts," but the number varies between worm species. Worms also have red blood that contains hemoglobin. Small blood vessels (capillaries) connect the different body parts to the main vascular network and not only bring nutrients and oxygen to the worm's body, but also remove wastes.

In earthworms there really isn't a brain, just a mass of neurons called a ganglion. This cerebral ganglion is connected to a pair of longitudinal nerve cords running the length of the worm's body. In each segment there is another pair of ganglia that are connected to the longitudinal nerve cords. Nerve fibers run from the ganglia and extend to the rest of each segment. On the ends of these nerve fibers on the skin, the sensory organs and cells can be found. These sensory organs tell the earthworm about its environment. The photoreceptor organs can sense changes in light intensity, and the epithelial sense organs can tell the worm if it's being touched.

Worms do not have lungs (though some of the aquatic species of annelid worms do have gills). They bring oxygen into their bodies by dissolving the oxygen through the body surface, which is kept moist by the mucus glands. There is a network of small blood vessels in the body wall that picks up this dissolved oxygen and carries it throughout the worm's body.

Earthworms need a lot of water in their environment. Not only do they need it to help keep them moist, so they can take in oxygen, but to replace large quantities lost through urination. One earthworm can produce 60 percent of its body weight per day in urine.

SEXING A WORM: ARE THEY MALE OR FEMALE?

Actually, they are both! Our friend the earthworm has both male and female reproductive organs, making them hermaphroditic. In *Lumbricus,* there are two male segments and one female segment.

When an earthworm matures in three to six weeks after hatching, the clitellum is formed to produce mucus for copulation, to secrete the wall of the cocoon, and to secrete albumin, in which the eggs are deposited in the cocoon. In the clitellum there are three layers of glands that perform these three different functions.

To mate, one earthworm will position itself pointing one direction while another will position itself pointing the opposite direction, so the head of one lies next to the tail of the other. The worms will lie close together and anchor themselves together by the longer setae on their reproductive segments. The clitellum of each worm secretes a mucus coat around the two worms, like a collar, further holding them in place.

In some worms, the male and female pores will line up, but

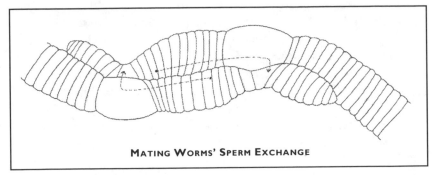

MATING WORMS' SPERM EXCHANGE

in *Lumbricus* the reproductive pores do not line up. Instead, the semen must travel a considerable distance from the male pore to the female pore. To accomplish this, muscles in the body wall of the segments contract and form a pair of sperm grooves. The groove is covered by the enveloping mucus layer secreted by the clitellum and thus becomes an enclosed channel.

The semen moves down the channel, carried by contractions of the muscles that produce the channel. When the semen reaches the seminal receptacles, the semen is passed to the other worm and taken into the receptacles. This process may or may not happen simultaneously in both members of the mating pair. Usually copulation or mating takes place over two to three hours and then the worms break apart.

A few days after mating, the worm secretes a cocoon in which the eggs will be deposited. To produce a cocoon, a mucus tube is secreted around the anterior segments, including the clitellum. The clitellum will then secrete a tough chitin-like material that encircles the clitellum. This will become the cocoon. The clitellum's glandular cells then secrete albumin for the eggs in the space between the clitellum and the tubular cocoon.

When all of this has been accomplished, the tubular cocoon will slip forward toward the front part of the worm. The eggs are discharged from the female gonopores, and then the sperm are deposited in the cocoon as it passes over the seminal receptacles. As the cocoon slips over the head of the worm, the mucus tube quickly disintegrates and the ends seal themselves, forming a completed cocoon. Cocoons can contain various numbers of eggs, from one to twenty, depending on the species; but, in *Lumbricus,* usually only one or two eggs per cocoon hatch. Adult worms may mate and produce cocoons continually every three to four days, throughout the spring and again in the fall months, depending on outside conditions. Worms kept inside in constant warm temperatures can reproduce throughout the year.

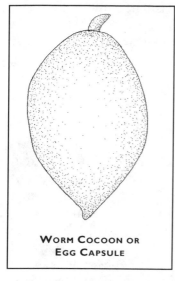

WORM COCOON OR EGG CAPSULE

Fresh cocoons are yellowish in color and look like tiny lemons. The cocoons gradually become darker as the embryo grows, feeding on the albumin deposited in the cocoon. Finally the young worms hatch from the ends of the cocoons. The length of time for cocoons to hatch varies greatly among species and depending on climatic conditions.

Experienced worm growers can double a population of *Eisenia fetida,* a popular composting worm, in just sixty to ninety days, so, as you can see, the reproductive potential of this worm can be quite high.

WHERE ARE WORMS FOUND?

Earthworms are found in all regions of the world now, except in deserts and frozen Arctic areas. They can be found in almost all soil types, provided adequate moisture and food are available.

As we discussed earlier, earthworms need moisture in the soil in order to breathe. The moisture in the soil, along with the mucus layer of the worm, allows oxygen to dissolve and pass into the worm. Earthworms can be found in soils containing as much as 70 percent water, but most consider a soil moisture content of 35 to 45 percent to be ideal. A common worm myth is that when it rains earthworms come out of their burrows to keep from drowning. Well, there are several possible reasons for this behavior—and none of them deal with drowning. One reason is that worms come out of their burrows when it rains so they can find a mate. Another one is that CO_2 levels in the burrow build up due to respiration, forming a weak acid solution that the worms do not like. Whatever the reason, studies have shown that worms can remain alive in aerated water. Fish breeders who feed worms to their fish report that worms can live for many months under the filter trays in aquariums. A bigger danger to worms is drying out.

The preferred diet of the earthworm consists of decomposing plant or animal matter, bacteria, fungi, and nematodes. Earthworms will even eat the dead and decaying parts of live plants, leaving the healthy parts alone. In many countries, part of the accepted agricultural practice is to leave cut plant material on the soil for earthworms to eat. This allows the earthworms to naturally fertilize the soil before the next planting.

A WORM BY ANY OTHER NAME

In this chapter we will be taking a closer look at individual earthworm species and earthworms in general. The study of worms is called oligochaetelogy. The "oligochaete" part comes from the Latin scientific name for earthworms, meaning "few setae" (bristles), and the "logy" part is from the Latin for "to study." You will notice that we will use the scientific name of a worm along with its common name. Most people resist using a scientific name, but when discussing individual worm species it becomes necessary. For example, the name "redworm" can refer to many different worms. You may think you understand which redworm you are ordering, only to find out when it arrives that it wasn't the worm you thought you were ordering at all. So, to avoid confusion, we will also use the unique scientific name of each worm species.

How does a species get a scientific name?

In 1758 a Swedish biologist named Carolus Linnaeus published a book on animal classification called *Systema Naturae*. Included in this book were more than four thousand animals, including man. Linnaeus was the first to give the scientific name *Homo sapiens* to man.

Linnaeus's system works like this: All animals and plants belong to groups that are similar. The first group, known as kingdom, includes all animals or all plants together and is the most general of all the groups. As the groups progress down from kingdom through phylum, class, order, family, genus, and species, the animals become more and more similar in appearance and behavior. The final two groups in the Linnaean system are genus and species. A genus contains organisms of similar characteristics and a species contains organisms that can interbreed.

In giving a scientific name to an organism, the genus and species names are used together. Our example of the redworm then becomes: *Lumbricus rubellus*. *Lumbricus* is the genus name, which is always capitalized, and *rubellus* is the species name, which is not capitalized.

WORMS IN GENERAL

There are thousands and thousands of different species of what we would call a worm. Anything that is long and legless, we term a worm or wormlike (for example, the worm lizard). There are nematodes, flatworms, leeches, tubifex, maggots (which are insect larvae), polychaetes or bristle worms, lug-worms, fanworms, bamboo worms, horsehair worms, and the list goes on and on. Worms can be found in marine, freshwater, or terrestrial environments. They can be free-living or sedentary. Some construct elaborate tubes and burrows. Others are parasitic and harmful to man.

Aristotle called earthworms nature's plows and the intestines of the earth.

In the phylum Annelida, or segmented worms, to which the earthworms belong, there are approximately nine thousand species. A vast majority of these worm species are aquatic, though we often think of the segmented worms as just the terrestrial earthworms we are familiar with. However, when we look at all the different worms in the soil, it is good to remember that earthworms and their relatives make up only a small number of the total worms present.

EARTHWORMS

The earthworm family of Lumbricidae, which includes the genera of *Lumbricus, Eisenia, Dendrobaena,* and *Allolobophora,* has hundreds of species of earthworms, but less than a dozen of these are important to cultivation. There are an additional twelve families besides Lumbricidae, in the class Oligochaeta, which are classified as earthworm cousins, and all together they total about six thousand species. There are two other classes of segmented worms, the polychaetes and the leeches, which make up the rest of the phylum. The worms can range in size from a few millimeters to the giant Australian earthworm, which can reach three meters. It would be impossible to outline the life histories of all earthworm species, so we will take a closer look at the few earthworms commonly used for vermicomposting and land improvement.

Terrestrial earthworms can generally be classified into one of three groups: the litter-dwellers, the shallow-soil dwellers, and the deep-burrowers.

The litter-dwellers live in the thin litter layer on the soil. In a forest, for example, you would find them just under fallen leaves or needles.

Shallow-dwelling worms, such as redworms, live primarily in the top twelve inches of soil. These worms do not build permanent burrows, but prefer to randomly burrow throughout the topsoil. When the weather gets colder in the winter, or the soil heats up and dries out in the summer, these worms will move deeper into the soil. Often you will find them down in the soil at about eighteen inches or so, rolled up into a mucus-covered ball. Shallow-dwelling worms may spend long summers and winters in this state of hibernation.

Deep-burrowing worms, like nightcrawlers, build permanent, vertical burrows that extend down into the soil six feet or so. Nightcrawlers are excellent soil aerators. Their burrows bring oxygen deep into the soil's top layer. Nightcrawlers are large worms reaching lengths of four to eight inches, and a few species can even reach twelve inches. A deep-burrowing worm will pull plant material down into the burrow, instead of burrowing through soil to find food. Sometimes the material is left just below the opening of the burrow to soften, and will be eaten later. Nightcrawlers are nocturnal, as their name suggests, and feed at night. Their feeding provides good soil mixing as well. When they pull decaying plant matter into their burrows, they mix it with soils from deep in the burrows. Finally, these soils will be deposited back on the surface with the worm's castings.

In areas where the land is constantly being turned over or

cultivated, you will not find very many nightcrawlers. These worms are more active in the spring and fall but do not go into hibernation like the shallow-dwelling worms in summer and winter. Instead they can retreat to the bottom of their burrows during temperature extremes.

NIGHTCRAWLERS
(Lumbricus terrestris)

COMMON NAMES: Nightcrawler, Dew worm, Night walker, Rain worm, Angle worm, Orchard worm, and Night lion.

COLOR: Red, brown, or a combination of these colors. Some have been known to be greenish.

ADULT LENGTH: Up to 12 inches

LIFE SPAN: Up to 10 years but generally only a year or two in the garden

HABITAT: Vertical tunnels that can be up to 6 feet deep (deep-dwellers)

DISPERSION: Widespread throughout Europe and North America. Some also found in New Zealand.

FOOD PREFERENCES: Leaf litter and mulch

TEMPERATURE: They like temperatures around 50°F (10°C).

COCOON INCUBATION TIME: 14 to 21 days

Nightcrawlers can regenerate lost parts of themselves, an ability that varies widely among worm species. Nightcrawlers have a relatively poor regenerative ability, but they have been known to do it. The closer to the middle the worm is cut, the more likely a successful regeneration will occur. However, sometimes a worm gets confused and regenerates a worm with two tails or two heads.

Nightcrawlers are not good for indoor vermicomposting systems. They like their burrows undisturbed and prefer to eat things that are found on the top of the soil. This can cause some problems with composting systems. However, they are a very important organism in nature for land improvement.

African Nightcrawlers
(Eudrilus engeniae)

COMMON NAMES: African nightcrawler, Giant nightcrawler

COLOR: Reddish with cream striping

ADULT LENGTH: Large—up to 12 inches

HABITAT: Top few inches, under litter and mulch (shallow dweller)

FOOD PREFERENCES: Rich compost

TEMPERATURES: 59 to 77°F (15 to 25°C)

The African nightcrawler can be a good worm for vermicomposting but tends to be restless. One day they are doing fine in the bin, and the next they are moving out. They require warm temperatures and are not recommended in areas that dip below 50°F unless kept inside. More work is being done to determine just how good a composting worm the African nightcrawler can be.

Redworms
(Lumbricus rubellus)

COMMON NAMES: Red worm, Blood worm, Red wiggler

COLOR: Somewhat iridescent on top, dark red to maroon. Lacks striping between segments and has a light yellow underside.

ADULT LENGTH: Up to 3 inches and has 95 to 120 segments

CLITELLUM: Covering segments 27 to 32, usually raised on top

FIRST DORSAL PORE LOCATION: Between segments 7 and 8

HABITAT: Prefers the top 6 to 12 inches of soil

FOOD PREFERENCES: Rich compost and decaying plant and animal material

TEMPERATURES: 64 to 72°F (18 to 23°C)

COCOON HATCHING: 12 to 16 weeks

Lumbricus rubellus is a very active wiggler in the presence of light. It is said that this worm is irresistible to fish and makes great bait because the worms exude amino acids that fish lack.

Lumbricus rubellus makes a good compost worm. Like nightcrawlers, they will aerate and mix the soil. They can be found in soils that have a rich organic component, such as animal pastures and compost piles.

RED WIGGLERS
(Eisenia fetida)

COMMON NAMES: Tiger worm, Garlic worm, Manure worm, Brandling worm

COLOR: Rust brown. There is a membrane between each segment with no pigment, and on each segment there are alternating bands of yellow and maroon down the length of the body.

ADULT LENGTH: Up to 3 inches

CLITELLUM: Covering segments 26 to 32 and raised all around the worm

FIRST DORSAL PORE LOCATION: Between segments 4 and 5

HABITAT: First few inches of soil (shallow-dweller)

FOOD PREFERENCES: Very rich compost, manure piles, and decaying plant and animal material

TEMPERATURES: 59 to 77° F (15 to 25°C)

COCOON HATCHING: Between 35 and 70 days depending on conditions

The red wiggler is an excellent vermicomposting worm. It can process large amounts of organic matter and in perfect conditions can eat its body weight in food each day. It also has a high reproductive ability and can double its numbers in sixty to ninety days. *Eisenia fetida* isn't too fussy about living conditions in the bin. It can tolerate fluctuations in temperature, acidity, and moisture levels that many worm species cannot. This worm also has some regenerative ability.

Eisenia fetida is used as a fishing worm, and, like *Eisenia andrei,* it exudes foul smelling coelomic fluid. (The Latin word *fetida* actually means stinky or smelly.) Some fishermen say that certain species of fish are attracted by this fluid, while others say the fish are attracted by the worms' wiggling.

In nature, these worms need soils that are extremely high in organic matter; they just cannot live in common garden or lawn soils. Of course there are exceptions to every rule, and some soils are very rich in organic matter. But think twice about adding extra *Eisenia fetida* worms to your garden without adding extra organic matter, because chances are they won't survive.

~~~~~~

RED TIGER
(Eisenia andrei)

COMMON NAMES: Tiger worm, Red tiger worm, Red tiger hybrids

COLOR: Dark red or purple. There is some disagreement over whether these worms are banded or not. Some worms have been identified with yellow bands between the segments, whereas others have been identified without banding.

ADULT LENGTH: Up to 3 inches

CLITELLUM: Covering segments 26 to 32 and raised all around the worm

HABITAT: First few inches of the soil and under mulch

FOOD PREFERENCES: Manure, rich compost, and decaying plant and animal material

TEMPERATURES: 64 to 72°F (18 to 23°C)

Eisenia andrei is a good worm for vermicomposting. It is a close relative of *Eisenia fetida* and also has the ability to process large amounts of organic matter.

These worms, like *Eisenia fetida,* are used for bait and exude coelomic fluid. They are very active wigglers in sunlight.

BLUE WORMS
(Perionyx excavitus)

COMMON NAMES: Blue worm, Indian blue, and Malaysian blue

COLOR: Anterior is a deep purple, while posterior is a dark red to brown. The clitellum and underside are light yellow.

ADULT LENGTH: Up to 6 inches

CLITELLUM: Covers segments 7 to 10 and is not raised. Some are even depressed.

HABITAT: Lives just under mulch (litter dweller)

FOOD PREFERENCES: Compost, decaying plant or animal material

TEMPERATURE: 68 to 77°F (20 to 25°C)

Blue worms are very active wigglers and make good fishing worms. It is a good vermicomposting worm in warm climates. It does not like cold weather and would not do well outside in cold regions. When this worm is used in indoor vermicompost, sometimes it will leave the bin for no reason. Blue worms fluoresce when exposed to sunlight.

Perionyx excavitus also has an excellent regenerative capacity and can regenerate any part it has lost.

SPENCERIELLA SPECIES

COMMON NAMES: Blue Worm, Indian blue, and Malaysian blue

COLOR: Deep purple on top with a dark red to brown underside. Clitellum is usually a light yellow.

ADULT LENGTH: Up to 6 inches

CLITELLUM: Covering segments 7 to 10, not raised

HABITAT: Top few inches of soil and under mulch (shallow-dweller). Australian native.

FOOD PREFERENCES: Compost and decaying plant and animal material

TEMPERATURE: 68 to 80°F (20 to 27°C)

This worm looks identical to *Perionyx excavitus* and is often confused with it. A very active wiggler that fluoresces under sunlight, they make good fishing worms and, unlike *Eisenia fetida*, reproduce by self-fertilization (parthenogenesis). They are prolific breeders and increase their numbers quickly.

3

How Do You Get Started?

Worms can be grown almost anywhere, from small Styrofoam ice chests to old refrigerators to large bins in outdoor buildings. In this chapter we will deal with small, home worm-bin composting systems that you can use to recycle kitchen wastes. We will take a look at larger, commercial worm composting systems in chapter 9, "Earthworms in Agriculture," and in chapter 11, "Commercial Worm Growing." Backyard worm beds for composting will be discussed in chapter 7, "Using Worms, Castings, and Vermicompost in the Garden."

WHAT IS A WORM BIN?

A worm bin is basically nothing more than a contained compost pile to which you have added earthworms. A worm bin can be made from wood, plastic, metal, or Styrofoam. As long as it conserves moisture and provides darkness for the worms, it can work as a worm bin.

In this chapter we will look at many worm bin designs. One is bound to fit your requirements and style.

What goes on in a worm bin?

Your goal for your worm bin is to put waste in and get (vermi)compost out, thereby recycling the nutrients. To do this, a complex series of events must take place.

In a compost pile without worms, the series of consumers goes something like this (assuming the environmental conditions such as air and moisture levels are right):

1. When organic material is placed in the pile, the first organisms to move in are the psychrophilic bacteria. These bacteria, which prefer temperatures below 70°F, start eating the fresh material. Over 70°F the mesophilic bacteria, which like to live in temperatures between 70 and 113°F, start taking over. Unfortunately for the mesophilic bacteria, their feeding and respiration causes carbon dioxide (CO_2) to be released and energy in the form of heat to warm up the pile. These conditions soon start killing the mesophilic bacteria.

2. The next organisms, the thermophilic microorganisms that love temperatures of 113 to 170°F, now move in and start consuming the organic matter and the dead mesophilic bacteria. This feeding continues until the food supply gets low, reducing the numbers of thermophilic microorganisms. The pile starts to cool down.

3. When the pile cools enough (70 to 75°F is ideal), the actinomycetes and fungi are ready to take over and eat whatever is left, including the thermophilic microorganisms. The actinomycetes are higher forms of bacteria that contribute to the formation of humus. They also improve the soil by releasing various nutrients, such as nitrogen and carbon. Actinomycetes are the organisms you smell in finished, earthy compost. The fungi are there too, eating their part. Fungi are primitive plants that do not contain chlorophyll and cannot produce their own carbohydrates. They must get their energy from the organic material by breaking it down into its constituent building-block compounds.

Now, if this was a simple compost pile, you could at this point add new organic matter and start the whole process again. However, in a vermicompost system, we add earthworms to our system and it changes the series of events a bit.

First we will consider the events and organisms we have just mentioned as first-level consumers. They are the first ones at the dinner table, so to speak.

In a bin system, we add earthworms as the first-level consumers, and try to discourage the rapid buildup of mesophilic microorganisms that produce high temperatures, which are harmful to earthworms. We try to keep a balance of food and worms by placing in the bins just the quantity of food the worms can eat. In most worm bins, the potential for heating does still exist, however, which is why certain foods that can result in the production of extreme heat are cautioned against.

In nature, earthworms will not move into a compost pile until after the thermophilic microorganisms have finished. Then the microorganisms in the pile, as well as the organic debris, become food for the earthworm. Of course, other organisms, such as mold, mites, springtails, grubs, and other bacteria, also join in to eat this large supply of food. These can found in vermicomposting systems, too. Some of these organisms worm growers consider pests, but many are microscopic and go unnoticed.

In a closed worm bin system, we keep repeating this process until we get the rich vermicompost we can use in our gardens. In an outdoor compost pile, something always eats something else and the pile will have many more levels of consumers.

First-level consumers	Second-level consumers	Third-level consumers
BACTERIA	SOW BUGS	CENTIPEDES
ENCHYTRAEIDS	MOLD MITES	PREDACEOUS BEETLES
FLY LARVAE	SPRINGTAILS	ANTS
FUNGI	EARTHWORMS*	PREDATORY MITES
ACTINOMYCETES	GRUBS	
EARTHWORMS		

*Earthworms and some other organisms can change consumer levels depending on what they are eating. When an earthworm eats food directly, it is a first-level consumer, but when it eats the bacteria in a compost pile after the initial heating process, it becomes a second-level consumer. For this reason, the old idea of a food chain was replaced by a more realistic food web.

What size bin do you need?

In order to decide how big a worm bin you need, you first need to determine how much and what kind of wastes you want to compost. Do you just want to compost your kitchen scraps, or do you intend to add yard wastes as well?

First, collect and weigh the wastes that you want to recycle for one week. **The basic rule of thumb is: for every pound of waste per week, you need one square foot of surface area for your worm bin.** So if your family only produces one pound of waste in an entire week, a worm bin to take care of this waste would need to be one foot high by one foot long by one foot wide. You could have a very successful vermicomposting bin for your family in a small Styrofoam ice chest if you don't produce much waste. Of course, no family produces exactly the same amount of waste every week, so taking an average over several weeks is even better.

Think of worm bins as these 1 x 1 x 1-foot cubes. If your family needs two square feet of worm bin area, place the cubes side by side, not one on top of the other. Remember, it is the surface area that is important. Cubes stacked on top of one another still only have one square foot of surface area instead of two.

Some people who want to recycle their kitchen wastes have their hearts set on one of those new manufactured worm bins. However, they don't produce enough kitchen wastes for that large a bin and the worms that come with it. In such a case they must feed their worms additional food. This could be in the form of yard wastes that have been partially composted first, purchased commercial worm food, or food purchased from the grocery store.

TYPES OF BINS

Commercial worm bins

Several different worm bins are now available for those people who are not interested in building one themselves. Most are made out of plastic and are lightweight and easily moved. Several come with bedding material and worms—all you add is food. This makes starting a vermicomposting system very easy.

Some beginning worm growers find these complete systems

easiest because everything is spelled out for them. However, some can be quite expensive. So before you buy one, do your homework and decide which one is best in your situation.

Tray system bins are becoming popular. These bins use vertically stacked trays for the composting system. You feed the bottom tray, and, when it is half-full of compost, you start feeding the second tray. This process continues until the top tray is full of compost. Then the first tray is ready for harvesting. This tray then goes on top to start feeding, and the process continues.

There are several plastic-container-type commercial bins, some of which come complete with worms, bedding, and instructions. They are light and portable and usually can handle about two to three pounds of waste per week.

In areas with severe rodent problems, metal bins might be the answer. They are made with twenty-six-gauge galvanized steel, and are extremely strong and rodent resistant.

Another option is to buy bin inserts. Inserts fit into name-brand plastic storage bins of the type that can be purchased at home centers and department stores. The inserts convert the inexpensive plastic container into an effective worm bin. The inserts cost less than a commercial worm bin, fit various sized containers, and are easy to install.

If a large-scale operation is what you have in mind, then you should look for a commercial, automated vermicomposting machine. This is a fairly large piece of equipment, not meant for household use.

In the buying guide at the end of this book, you'll find a small list of distributors of all kinds of commercial worm bins and inserts. Most retailers will be happy to send you a price list and more information concerning their products. Check with your local recycler or home center to see if they carry worm bins. Many nurseries and garden centers give demonstrations on how to set up a worm bin.

Making your own worm bin

Any container can be turned into a worm bin. Successful worm bins have been made from picnic coolers, Styrofoam ice chests, and even old refrigerators. All three of these containers have excellent insulation and can keep worms quite comfortable.

PLASTIC WORM BINS

A plastic storage bin purchased at a home center can be made into an excellent worm bin. Plastic does not breathe, so extra holes must be added to the sides and bottom for aeration and drainage. One type of bin provides extra space in the bottom of the bin for liquids to collect, so holes don't have to be drilled for drainage. Also, plastic is not a good insulator. Plastic bins are most often used indoors where temperatures remain fairly constant.

EXAMPLE #1:

MATERIALS NEEDED

- One medium (14 to 18-gallon) plastic bin with lid
- Landscape shade cloth to fit bottom and sides

INSTRUCTIONS

Drill six 1-inch holes in various locations in the long sides of the bin and four holes in the short sides. Next, drill four holes in the bottom of the bin. These holes will be used for drainage and aeration. If the bin becomes too wet, you can drill a few more holes later. Bins in dry climates may need fewer holes than bins in humid regions.

Clean bin with soap and water. Dry.

> NOTE: Old plastic containers can also be used for worm bins. These should be free of chemicals and dirt that can harm your worms.

Add the landscape cloth. Cut a piece that will fit on the bottom and a long piece for around the sides. The cloth will prevent bedding and casting from falling out of the holes. You can also cut squares of landscape cloth and hot-glue them over the holes.

That's about it! Use the lid as a bottom tray to catch worm tea, and use leftover landscape cloth as a lid (or buy an extra lid if you can). Add bedding and worms.

MATERIALS NEEDED

- One 18-gallon plastic bin with lid
- One 4 $^1/_2$ foot-long 1" x 4" piece of board lumber (not pressure treated)
- One 12" x 18" piece of window screening
- Staples
- Sphagnum peat moss

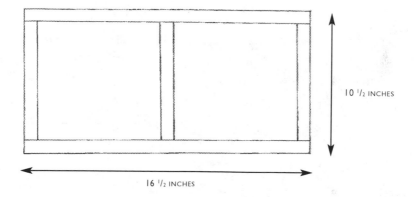

10 $^1/_2$ INCHES

16 $^1/_2$ INCHES

FRAME DIAGRAM: EXAMPLE #2

INSTRUCTIONS

Cut the board into two pieces that are 10$^1/_2$ inches long and two pieces that are 16$^1/_2$ inches long. Use these pieces to construct a frame. The 10$^1/_2$-inch pieces are the ends and the 16$^1/_2$-inch pieces are the sides. Make a brace across the middle from left-over pieces of board.

Place the screen over the frame and staple it on securely, then place the frame with screen-side up in the plastic bin. Add dry sphagnum moss until the top of the frame is covered with about two inches of moss. There will be an empty space on the bottom now, under the frame and screen.

Now add about eight more inches of moist bedding on top of the dry moss. This can be more moss or whatever bedding you prefer. Cut a two-inch hole in the lid for ventilation and cover it with screen. Use a hot glue gun or tape to attach the screen.

This plastic bin does not require drilling any holes. The liquid, if any, will collect at the bottom in the empty space and can be poured off when it is time to harvest the worms. Since there is a lot of dry material in this bin, the moisture levels should be checked to prevent the bin from becoming too dry. Generally, in plastic bins that won't be a problem.

EXAMPLE #3:

There are some people who are converting old nursery flats into worm bins. The bin is made like this:

MATERIALS NEEDED

- Four plastic nursery flats or trays with holes in the bottom
- A piece of heavy plastic bigger than the width of the tray
- Fifteen small wood blocks or a $1\frac{1}{2}$-inch-wide dowel, cut into pieces $\frac{1}{2}$ inch shorter than the depth of the nursery trays. These will support the corners and centers of the trays.

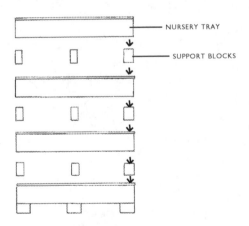

NURSERY TRAY

SUPPORT BLOCKS

NURSERY TRAY SYSTEM: EXPANDED VIEW

INSTRUCTIONS

Line one tray with heavy plastic. The plastic should come up the sides and over the top of the tray. This will be the bottom liquid collection tray. If you can get a nursery flat without holes, use it for this layer.

Lay the solid or plastic-lined tray down first. Place blocks in the corners and one in the center. Add the next tray (Tray #2). Place blocks in the corners and center. Bed this layer with an inch or two of bedding and add food and worms. This bin is small—use about ½ pound of worms to start. Every time you feed the worms, add about one inch of damp newspaper, leaves, coir, or your favorite bedding to the tray. When the worms are about to the top of this first tray, add the next tray (Tray #3).

Tray #3 now should be bedded and the worms fed there. Do not feed Tray #2 any more. The worms will move up through the holes to the fresh bedding and food, leaving the lower tray. Continue this process right up to the last and top tray (Tray #4). Then take the bottom tray with holes, Tray #2, and empty the vermicompost and place it on top. This becomes a continual cycle of trays.

Empty the bottom tray, Tray #1, of liquid when needed.

NOTE: If you prefer stronger trays, try building the trays out of 1" x 4" lumber and using ¼-inch hardware cloth, stapled to the bottoms. Make the bottom tray solid.

WOODEN WORM BINS

Wood is an excellent material to use for worms. It breathes better than plastic and has better insulating qualities. Wood does have some drawbacks: it is heavy and it rots. So, wherever a wood worm bin goes, that's pretty much where it stays.

When purchasing wood to build a worm bin, do not buy pressure-treated lumber. Many organic gardeners do not put pressure-treated lumber in their gardens because the toxic chemicals used as preservatives can leach out of the wood. These chemicals (copper, chromium, and arsenic) can also be toxic to your worms. Cedar is also not recommended for bin construction.

Untreated wood unfortunately will last only two to three years with continuous use. Worm bins are damp environments and the moisture helps to break down the wood fibers faster.

Some worm growers use two wooden worm bins and rotate the worms between the bins to allow the wood to dry out a bit, making it last longer. Another idea is to coat the inside of the bin with cooking oil.

A simple wooden box, like a dresser drawer, can also be used as a worm bin. Actually, worm bins should be pretty shallow. Redworms are by nature shallow feeders and you don't need a deep box. It's the surface area that is really important.

Note: When dealing with finished lumber, keep in mind that when you ask for a 2" x 2" piece of lumber, you will get a 1½" x 1½" piece of lumber. (Measure your pieces before you go home to double-check the dimensions.) Some lumberyards may have unfinished pieces you may want to use; these will be a different size. Lumberyard employees are generally very happy to help with measurements, so take the directions with you when shopping. They may even have scrap pieces laying around that they can give you. The old carpenter's motto applies to any woodworking project: Measure twice, cut once!

EXAMPLE #1:

MATERIALS NEEDED

- One 7-foot length of 2" x 2" lumber
- One 8-foot length of ½" x 12" board lumber
  (for a bin with no lid)

or

- One 10-foot length of ½" x 12" board lumber
  (for a bin with a lid)
- Thirty-five to forty ½" screws

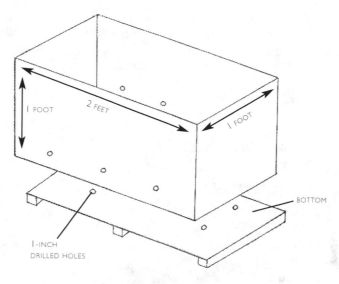

BIN DIAGRAM #1

INSTRUCTIONS

Cut 2" x 2" into: Four 1-foot lengths (inside corner supports)

Three 1-foot lengths (bottom support)

Cut the ¹⁄₂" x 12" board into: Two 11-inch-long pieces (short sides)

Three 2-foot-long pieces (long sides and bottom)

Assemble the box frame as shown in diagram #2.

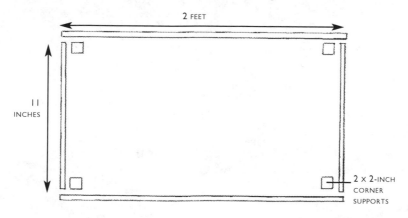

DIAGRAM #2: EXPANDED TOP VIEW OF BIN AND CORNER SUPPORTS

Screw all the pieces together. Place the last 2-foot piece on the top of the box. This piece will form the bottom later. Now place the remaining three, 1-foot 2" x 2"s, one on each end and one in the middle. Screw together, attaching the bottom to the sides.

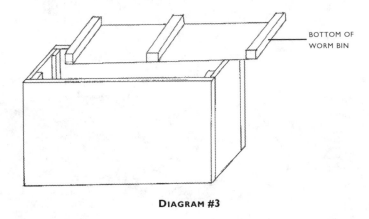

BOTTOM OF WORM BIN

DIAGRAM #3

Finally, drill 1-inch holes in the box, four holes in the bottom and six holes (three at the top and three near the bottom) on the two long sides.

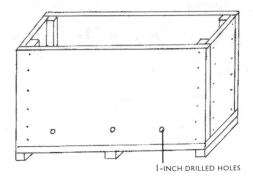

1-INCH DRILLED HOLES

DIAGRAM #4

Add bedding and worms.

To add a lid to this box, simply cut another piece of board 2 feet long and attach it with hinges. You can add a frame of 2" x 2" to the inside of the lid to stabilize it.

Remember, this box will drain. Place two shallow pans under the holes to catch the worm tea. Empty as needed.

This worm bin is a small one with only two square feet of surface area. It would be great for a single person or couple or a family that only produces a small amount of waste per week.

Worm bins can be expanded depending on your needs. Using the same idea, a bin measuring 2 inch wide by 3inches long by 1 inch high could be constructed. For this you would need:

EXAMPLE #2:

MATERIALS NEEDED

- Two 8-foot lengths of $1/2$" x 12" board lumber (and another 6 feet, if you want a lid)
- 13 feet of 2" x 2" lumber (Buy two or three pieces of this to equal 13 feet, if needed. These pieces can be pieced together easily.)
- 1 $1/2$-inch screws

INSTRUCTIONS

Make the sides of the box just like for the bin in Example #1, except the short side pieces should be 1 foot, 11 inches and the long side pieces should measure 3 feet. For the bottom, cut two pieces of lumber 3 feet long.

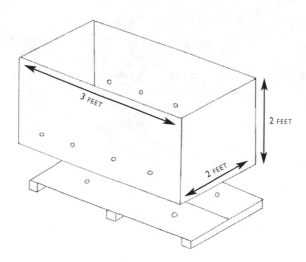

3 FEET

2 FEET

2 FEET

DIAGRAM #5

Attach the bottom to the sides, like in Example #1. Drill holes.

Make lid, if desired, by using two 3-foot-long board pieces. Place them side by side and join them together with brackets or extra boards. Attach to the bin with hinges.

This bin could also be made from 1/2" to 5/8" exterior grade plywood. The nice thing about plywood is it is stronger than regular board lumber and there are no seams to piece together. But the drawback is that it comes in large sheets—usually 4 x 8 feet sheets—and requires cutting with a circular saw.

NOTE: Use the same principle to make an even larger bin if needed. Maybe 3 feet wide by 4 feet long by 1 foot deep? Sometimes it is easier to make several smaller bins, to make it easier to harvest and clean them. Also, small bins have an advantage when it comes to moving them, as they are not impossible to move. And, if the worst happens and something goes wrong in one bin, all is not lost.

Some worm growers suggest covering air holes with screening material to keep unwanted pests out of your bin. This is good advice for outdoor bins. If flies are a problem in your area, this is a good idea for indoor bins, too. Window screening or landscape shade cloth can be cut, stapled, or glued over the holes.

Larger and more elaborate worm bins are starting to evolve as more people explore vermicomposting. There are those who have turned patio storage seats, coffee tables, and flower planters into worm bins. With a little imagination, almost any container that has a pretty large top surface area can be transformed into a worm bin. Even old pieces of furniture, like a child's old wooden toy box, can be converted into worm bins. Generally, all that is necessary is to add aeration holes and elevate the box off the ground. Check carefully when using old furniture that it is not covered in toxic paint. If you are not sure, sand the paint off the inside to bare wood or use a converted plastic storage bin inside. The plastic bin will keep the moisture from destroying a lovely wood piece and make it easy to simply lift the bin out and clean it outside.

Worm bins can also be constructed with mesh bottoms. This allows a lot of aeration and lots of drainage for bins kept outside. Moisture levels should be checked often with mesh-bottomed bins to prevent them from drying out. In the winter, an insulating layer of newspaper or cardboard at the bottom of the bin will help keep the worms warm. If the bin rests on the ground, the open mesh will allow microorganisms that assist in the breaking down of organic material easy access to your bin.

HOW MANY WORMS DO YOU NEED?

Worms typically eat an amount equivalent to approximately half of their body weight per day in good conditions. Once you have decided how much total waste you have to feed your worms in an average week, divide that number by seven to get the average amount of food you have for worms each day.

A simple example using this formula is:

1. A family produces seven pounds of total waste in an average week.
2. Divide the seven pounds of waste by seven days in a week.
3. The answer is one pound of waste produced each day to feed the worms.

So, knowing this, our example family needs two pounds of worms to eat the one pound of waste they produce each day. Of course, this is an average; when conditions in this bin are good, worms will eat more, and when conditions aren't so good, worms will eat less. In chapter 4, "Maintaining A Worm Bin," we will go into a more detailed explanation of conditions and feeding worms in a worm bin.

Now, your next question is: How many worms are there in a pound? Well, it really doesn't matter, because commercial worm growers sell their worms by the pound. However, most worm growers estimate there are approximately 1,000 adult redworms in a pound.

WHERE TO GET WORMS

After you have figured out how many worms you need, it's time to buy the worms. If you purchase commercial bins, the correct number of worms comes with them, and you won't have to worry about this. However, for those who are building their own bins or those who have had a major problem and need to start over, you need to know where to buy and what you are buying.

Worms can be purchased in some garden centers and bait shops. The only problem with some of these outlets is that you usually have no idea what worm you are getting, beyond "redworms" or "nightcrawlers." Nightcrawlers are not used in worm bins because they construct permanent burrows, but you can use bait shop redworms. We recently went to many of the bait shops in our area and the only thing written on the containers was "redworms." We asked the bait store operators what kind of redworms we were buying, but they didn't know. These unknown redworms would work in a worm bin, but it would be nice to know exactly which worm you are using so you could adjust conditions in the bin to their liking. Another problem is the cost. Bait shop worms are very expensive.

There are commercial worm growers that will be happy to ship you all the worms you need, usually at a better price than retail outlets. Ask for the worms' scientific name, so you know exactly which worms you are ordering. Many commercial worm growers will ship worms in a few days, depending on the weather. It is best to order worms when they can be shipped in moderate weather; extremely hot or cold conditions are not the best for shipping worms. You can find many commercial worm growers that advertise in gardening publications or on the World Wide Web.

When you order worms, the grower may ask you if you prefer breeders or bed-run. Breeders are all mature worms that will generally cost you more than bed-run, which are mature and immature worms mixed together. Some people think breeders are best to start with because they will produce cocoons more quickly, but it won't take long for the immature worms to grow up, either. An advantage to having young worms is that they will adjust to their new home much faster than adults. If you are looking to receive higher numbers of worms, then buy a bed-run shipment by weight. The average estimate is 2,000 per pound for bed-run and 1,000 per pound for adult worms. Your can decide which is best in your situation.

If you are lucky enough to live close to ranch lands or people who have horses or livestock, you might try finding redworms yourself. Of course, get permission to do this first—a person with a shovel and bucket out in a field might get into some trouble.

Try turning over a few piles of decomposing manure. You might get lucky and find a jackpot of worms.

WHERE SHOULD YOU PUT
YOUR WORM BIN?

When deciding on a place to put a worm bin, do your homework concerning the living requirements of the worm species you will be raising. This is especially critical if you decide the worm bin must remain outdoors. In colder climates, outdoor worm bins may need to be insulated in the winter, and in hotter climates, the bin may need to be insulated in the summer. Your weather will dictate where your worm bin should go: in a shed, in a basement, or on a patio. Most composting worms work best at temperatures between 59 and 77°F (15 to 25°C). Worm growers agree that temperatures under 50°F or above 86°F can be harmful to worms.

Besides finding the right temperature for your worm bin, choose a place that is convenient for you to tend. Worm bins need water, so locating larger bins next to a hose outlet would be a good idea. Small bins may only need sprinkling from a watering can.

If you choose a large worm bin that cannot be moved easily, remember that changing the bedding and doing maintenance can at times be messy. An outdoor worm bin keeps the dirt and mess outside. Small bins can be cleaned and changed at a utility sink.

Many worm growers who grow worms outdoors or in sheds find that it is necessary to keep a light on over their worm bins to discourage worm migration. So it's good to have an electric outlet nearby, just in case.

Remember to have a place for storage of tools and items you need to take care of your worm bin.

Finally, locate the bin in a place that will be aesthetically pleasing for you. There are people who have worm bins as decorative coffee tables, but many people tend to keep their worms in the garage or basement, or on a patio. Some with very small bins can keep them under the sink.

Do not place a worm bin next to a working refrigerator or

anything that vibrates. Worms do not like a lot of moving around and will not do well in those circumstances.

Extra insulation

Worm bins that are kept outdoors may need extra insulation in the winter or summer months, depending on where you live. Some bins that are not too heavy can simply be moved inside a shed, barn, or basement, but many are too heavy to move.

There are just about as many ways to insulate a worm bin as people have ideas. The following is a list of insulation ideas that people are using today to keep their worms comfortable and happy.

- Cover the worm bin with Styrofoam or thermal insulation. It comes in large sheets and can easily be cut to desired lengths and widths. It can be added right on the existing box. So, when finished, the worm bin is completely covered. The insulation can be nailed or glued directly to the box. Styrofoam or thermal insulation does not breathe, so make sure the worms have some ventilation.

- Stack hay bales around the box. This can be as simple as stacking the hay around and over the box. When you need to tend the box, you will have to remove some of the bales. We have also seen whole hay bale sheds constructed just for worm bins. There is a lot of information about hay bale construction, so it is best to check out all the hay bale designs and decide which one is best for your worm bin.

- Locate the bin in a shady area. This is a simple and sometimes forgotten idea that can help when the weather is hot. Locate the bin under a large tree or porch or on the north side of a shed or building. Try placing bushes around your bin.

- Place a low-wattage night-light inside the box to bring up the temperature. Be sure to place several layers of thick newspaper between the light and the worms.

- Add a bit of fresh green waste to the bin. The thermogenic bacteria in the early composting process will really heat things up. You don't have to use much. Experiment with amounts and check occasionally with a thermometer. Place the waste on one side of the bin, so the worms can escape to the other side of the bin if it gets too hot.

- Place wet towels over the bin on really hot days. The evaporation will help cool the bin. Add a fan if needed for more air flow.

- Heaters made for reptiles and bird baths can also be used. Check local pet stores for ones that can be kept in a moist environment and can be adapted. Some come with thermostats that can also be adjusted.

Many of these ideas can be used in conjunction with each other. If you live in an area that is extremely cold or hot, you may need to use several of the ideas together.

Good luck!

WHAT TYPE OF BEDDING SHOULD YOU USE?

Among all the ingredients and materials necessary for a vermicomposting worm bin, bedding is one of the most important. Bedding usually refers to the loose, moisture-retaining organic material used to fill the worm bin or windrow. Bedding provides carbon for bacteria, which constitute the bulk of the nutrients in the worms' diet and also break down the organic material fed to worms so it can be ingested by worms and other organisms. The importance of good bedding material is one of the least understood aspects of vermicomposting, and improper maintenance of this bedding is one of the main reasons for the failures encountered by those who are new to the process.

When starting a worm bin, you must first consider what a worm needs to survive. First it needs food to eat, environmental conditions that do not change too rapidly, shelter from light, access to other worms for mating purposes, and delicate handling. But most worm growers first think of what they need: healthy and productive worms, an odor-free and pest-free bin, and cheap, easy to obtain materials.

Bedding has to be one of the biggest variables in any worm bin system. Worms can be placed in bedding that is too dry, too wet, too dense, and even too coarse, and all of these can lead to problems. Some worms end up living in their food, with no bedding at all. These situations can certainly make caring for your worms a lot more difficult.

All bedding materials should conform to certain requirements. Bedding should:

- retain moisture in a form that is accessible to worms;
- stay loose and allow for air passage between the individual pieces of bedding;
- allow for drainage of excess moisture;
- not be too coarse;
- not be a food source that is high in protein;
- be aged past the heating stage, for manures and green wastes; and
- be a carbon source for bacteria.

Even when a type of bedding meets all of these requirements, you should always test a small group of worms on the new bedding first. If the test group survives and is doing fine after twenty-four hours, there's a good chance that the rest of your worms will be fine, too. So, with all of these points in mind, let's look at several bedding options and the various pros and cons associated with each.

Peat moss

Peat moss, or sphagnum moss, has been one of the most widely used of all the worm beddings for commercial worm growers. Many people have preferred to use "Canadian" peat moss, because it is believed to be a much more sterile medium; whereas, American peat moss is more likely to contain impurities and be tough and stringy, which might prove harmful to the worms. However, some worm growers use American sphagnum moss and thoroughly "leach" it before using it as worm bedding. To leach the moss, it must be soaked in water for several hours and then squeezed of excess water. This is done several times until only clear water runs from the moss. Many breeders still prefer to ship worms in Canadian sphagnum moss.

Peat moss has advantages and disadvantages. However, many worm growers believe that, despite the advantages, the moss should not be used because it is a nonrenewable resource.

ADVANTAGES:

- Moisture retention. Because it retains moisture so well, peat moss can keep moisture levels up, so that even in dry climates the bin doesn't have to be watered as often.

- Cleanliness. This is one of those characteristics you can only appreciate if you have had prior experiences with bedding materials that are not so easily handled (such as manure). Properly moistened moss is easy to handle, and, if you should drop a little, just let it dry, and vacuum or sweep it up.

- It has no odor. Though worms will convert bedding and food to earthy-smelling vermicompost, sometimes a bedding can take on an odor that some people may find unpleasant. Peat moss will not do that.

- Availability. Most garden-supply shops or nurseries will have this material on hand. A medium-sized bag can usually be purchased for just a few dollars, but larger quantities can get quite expensive.

- Consistency. A large bag of moss can last several months for the average indoor worm system. This can prevent the worms from being subjected to the rapid changes that can occur when the bedding is changed.

- As an enhancer to other bedding. Sometimes a bedding will become dense and need loosening up, or sometimes the bedding will dry out too quickly. Peat moss is perfect for correcting these problems. By adding a 30 to 50 percent mixture of moss to the bin, the necessary moisture level can be retained. In beddings that are too dense, peat moss will open it up and allow more air into the bin. Peat moss is acidic, and when mixed with bedding materials that are on the alkaline side, the peat moss can serve to bring things into a better pH balance.

- The cost. Many types of worm bedding are free, and since most people must buy peat moss, this can be a disadvantage.

- It is acidic. Peat moss is slightly acidic, and worms sometimes will not like moss as a bedding if they were kept in an alkaline bedding before. You may need to gradually acclimate the worms to the new conditions. To do this, simply transfer some of the old bedding material, worms and all, to one side of the new bed, opposite the moss bedding. The worms will then move into their new bedding at their own pace. This procedure should be used whenever you place newly received worms into a bed that contains material other than that in which they were shipped.

- It's a nonrenewable resource. This means it cannot be reproduced, and is therefore recognized as environmentally and commercially unsustainable. This is peat moss's biggest disadvantage, and one reason many commercial worm growers no longer use or recommend it.

All in all, peat moss is truly a remarkable substance, and a bedding material once widely used by vermiculturists around the world.

Newspaper

Newspaper is a good bedding material, especially when mixed with dry leaves or straw, which prevents the paper from compacting. Black inks that are now used in newsprint are nontoxic to the worms, too.

ADVANTAGES:

- Generally cost-free. Instead of recycling the morning paper at the recycling center, use it in the bin.

- Readily available. You can obtain newspaper anywhere.

- Odorless.

- Easy to prepare. Just tear it into thin strips.
- Recycles the paper. Worms can turn the newspaper into a wonderful product.

- Torn newspaper tends to compact in the bin.
- Ink can rub off on your hands.

It should be mentioned that we are talking about regular newsprint, not the glossy, multicolored advertisements found in the newspaper. There is still some debate about how safe the colored inks are for worms, so these should probably be avoided.

If you have a paper shredder, then use it to shred your newsprint. The smaller pieces of paper will hold moisture better than larger hand-torn paper.

Many offices now shred documents for disposal, and there has been quite an argument about using these papers from computers and copiers in worm systems. There has been no research about whether using such paper will cause soil and compost contamination, so some would say it is better not to use them. But some worm growers say that the levels of toxins associated with these inks are so small they should not be a problem. So, the debate goes on. There are several worm growers using shredded computer paper and having no problems. If you have access to these papers, you might want to try a small test bin and see for yourself.

Compost from plant material

This type of bedding can be any plant material that has been broken down past the heating stage. Some worm growers like leaf mold (partially decomposed leaves), while others use only composted grass clippings.

ADVANTAGES:

- No cost. Many people are happy to find a place that will take this stuff so that it doesn't go to landfills.
- Compost is a worm's natural environment.

- Time. You must wait for the decomposition process to be far enough along (past the heating process) or you could cook your worms.
- Compost, especially leaf mold, can compact in the bin.
- It can bring in uninvited guests to your bin. Since you must compost outdoors, other composting organisms can be present in the compost.

Animal manures

Manure that has been partially composted is a perfect habitat for worms. It is high in organic compounds that composting worms love.

ADVANTAGES:

- In some areas it is free for the hauling. If you're lucky enough to live where livestock is kept, many ranchers would love to have you haul off as much as you want.
- Worms really love it. For some, it's their natural habitat.
- Manure contains high levels of nutrients. It can be a food stock, too.
- The castings that come from worms living in composted manure bedding are extra nutrient rich.
- Small amounts of uncomposted manure can be added to outside beds or bins. This manure can warm up a cool bed a bit and worms will nibble at the edges.

DISADVANTAGES:

- May not be available to everyone for free.
- The odor of composting manure can be very objectionable for the first few days.
- Many other organisms, like mites, centipedes, and flies, also like composted manure and can be brought into your worm system.

- Time. You must compost the manure past the heating stage. This can take from a few weeks to several months, depending on weather conditions.

- Some manures should be leached (washed) to rid the manure of de-worming medications, urine, and salts.*

- Manure can compact easily in the bin. Other animal manures also need to be washed, for various reasons. Horse manure needs to be washed to remove the de-worming medications that are routinely given to the animals. This medication will be harmful to your worms.

Rabbit manure usually contains a very high content of urine. Repeated washings are necessary to clean it thoroughly. It can also be very acidic; some worms growers who use rabbit manure will add calcium carbonate (limestone flour) to neutralize the acidity. Be sure to check with your pH meter to be sure.

Sheep, goat, and pig manures can also be used, but they have

*Washing manure:

Where dairy cows and steers are kept in close quarters, their manure has been shown to be high in urine and salts, so washing these chemicals out of the manure becomes necessary.

To wash the manure, you may need to soak it, so that water is able to completely pass through and dilute away any harmful chemicals. Some worm growers will wash their hauls of manure right in the truck bed. Just add water to the truck bed until full and then allow the water to drain out. Another way to do this is to add manure to a garbage can that has holes in the bottom. Fill the can halfway full of manure and then fill the can up with water. Allow the can to drain. Some worm growers recommend filling the can three to five times to be sure the manure is thoroughly washed and free of harmful chemicals.

Next, place the washed manure in piles to dry and compost. This process can take from two weeks to several months, depending on weather conditions. When the composted manure is ready after the heating process (always wait at least two more days after the heating has stopped before introducing worms), many growers will test the manure with a small group of worms. If the worms aren't adversely affected after twenty-four hours—that is, you don't see them dying or trying to escape—then the manure is probably safe for the rest of your worms.

a higher nitrogen and protein percentage than steer, horse, and rabbit manures. They should be completely composted, which will take a bit longer; they can be more acidic as well. Calcium carbonate may have to be added to these manures as well. Check pH carefully.

Poultry manures from chickens and turkeys have special problems. Heavy metals are associated with these manures, so they must be thoroughly washed. Most worm growers use only small quantities of poultry manures in feed stock and not in bedding, because of the very "hot" nature of this manure. It has a very high content of nitrogen and protein

Coir (coconut fiber)

Coir, also known as coia, is the shredded fiber of coconut husks. It has the consistency of peat moss and some of peat moss's great moisture-retaining properties. It is becoming more and more popular, since it is a renewable resource, unlike peat moss. For the areas that produce coconuts, the discovery of a use for coir is very beneficial economically.

ADVANTAGES:

- Retains moisture very well.
- Mixes well with other bedding material to help retain more moisture and to fluff up dense beddings.
- Odor-free.
- Very clean to use. Like peat moss, if you spill some, it's no big deal—just let it dry and sweep it up.
- Has a good pH of about 6 to 6.5.
- Doesn't compact in the bin.

DISADVANTAGES:

- Cost. It can be expensive, considering that a lot of bedding materials are free.

Wood chips

Wood chips can make an excellent bedding. Most worm growers mix wood chips with other bedding materials because they can dry out easily.

- Very clean.
- Odorless.
- Mixed with other bedding, they keep bedding open and loose and allow air to get in readily.
- They can be reused. In bedding, the wood chips aren't broken down quickly, and usually, when the worms are harvested, the wood chips can be picked or screened out and used again in another bin.
- Cost. Many people can obtain wood chips from landscapers for free.

DISADVANTAGES:

- Cost. Many people cannot get wood chips for free and must purchase them in bags.
- They dry out quickly.
- They are large and must be mixed with some other bedding.

Sawdust can also be used as a bedding, much like wood chips. People who do woodworking have a large supply of sawdust.

BEDDING THE BIN

There are various ways to do this, depending on the kind of bin you have and what type of bedding you choose, but we will explain a simple type of bed. If you have purchased a commercial worm bin, follow the instructions that came with the bin.

Once you have decided on the type of bedding, prepare it if necessary. Many of the most successful worm bins are a mixture of several beddings. Don't feel you have to use only one type of bedding.

Fill your bin three-quarters full with the bedding. Make sure the bedding is just damp; when you tightly squeeze the bedding, no more than a few drops of water should come out.

Add one to two handfuls of garden soil to provide grit for the worms and the necessary microorganisms for composting.

This is generally done for brand-new worm bins. For rebedding a worm bin, a handful or two of vermicompost will add the necessary composting microorganisms. Then you can add one of the additives below for grit. Mix a bit.

Add the worms, and you are off to the races.

Don't forget to feed the worms, too!

Bedding Additives

As we just mentioned, soil should be added to the bedding to provide composting bacteria and microorganisms that will help the worms turning wastes into vermicompost. There are other additives to neutralize the pH in the bedding (improving conditions for the worms) and to provide extra minerals for the worms, microorganisms, and finally for the plants.

Dried eggshells are perfect for neutralizing the pH of acidic beddings and for providing grit for the worms.

Calcium carbonate will do the same thing, but eggshells are free. If you do decide to purchase calcium carbonate (powdered limestone) remember that a little goes a long way. A word of caution: Hydrated lime is not the same thing as powdered limestone. Store clerks have gotten the two confused, but hydrated lime will make your worms very unhappy and in many cases kill them.

Rock dust is another product many worm growers use and recommend. It provides grit and minerals for the worms and microorganisms. The vermicompost produced with rock dust is full of trace minerals needed for good plant growth.

4

MAINTAINING A WORM BIN

WHAT TO FEED THE WORMS: DO'S AND DON'TS

Most organisms need a balanced diet, and earthworms are no exception. They prefer a balanced diet of cellulose (carbohydrates), fat, protein, and minerals. This diet sounds much like our own—and it is. The organic material that is fed to worms is usually referred to by worm growers as "feed stock." It is usually a nitrogen-rich material that also provides energy to the bacteria in the bin. In some cases the feed stock and the bedding material are the same, but this isn't always the case. Most feed stocks are very high in moisture and lacking in bulk. These foods soften quickly and don't provide enough space for oxygen to get down into the bedding material.

Most people start a home vermicomposting system so they can recycle the kitchen wastes they produce. Vegetable wastes provide a basic vegetarian diet that is perfect for worms in a home vermicomposting system. However, commercial worm food is also available. One advantage to commercial food is you know it is balanced and worms will eat it. Some worm growers say it is good to use commercial food when starting a new bin, so, if your worms are unhappy with their new home, you will know it is not the food they are objecting to. This sounds great, but commercial worm food isn't available everywhere; it is manufactured in Australia and sometimes the cost can be prohibitive. So what do you feed your worms?

Food examples

Vegetables (including peels and tops)

Fruit (peels and flesh)

Coffee grounds, including filter

Tea leaves and tea bags

Kelp meal

Plain breads

Rice

Cornmeal

Pasta

Cakes, muffins, biscuits

Melons (a worm favorite)

Brown sugar

Floral arrangements
(be sure the plants are not poisonous, like oleander)

Crushed eggshells

Cereal

Pizza crusts

Cheese (small amounts)

Macaroni salad

*Composted or aged green waste and manure from plant-
eating animals like horses, cows, rabbits, and sheep

And many more!

---

*We must emphasize here that green waste and manures must first be composted. The main reason for this is that green wastes can produce high temperatures and nitrogen-based gases that can be lethal to worms. Also, manure may contain high levels of salts and ammonia, which must first be leached out. For detailed instructions, see chapter 3.

FOODS TO AVOID OR LIMIT

Citrus

Meats and bones

Garlic

Heavily spiced foods such as many Asian and
 Mexican dishes

Hair

Dairy products: milk, yogurt, or butter

Eggs

Fresh green wastes and fresh manures

Poisonous plants

Oils

Salt

Wood ashes

Pet feces

Absolutely no metals, foils, plastics, chemicals
 (including solvents and insecticides), or soaps
 should be placed in the worm bin at any time.

Many of the foods listed in "Foods to avoid or limit" are not necessarily bad for worms. Foods such as meat and eggs are not recommended for home vermicomposting bins because they have some unpleasant side effects in the bin. The biggest problem is odor. Meat and dairy foods can produce very strong odors when the proteins in them break down. Most worms won't at all mind chewing on a nice chicken bone once in a while, but, for bins kept inside, care must be taken when feeding these foods.

Another problem associated with the decomposition of proteins is that it will attract other animals that can become pests. The smell of a nice rotting piece of meat is something a fly just can't resist. Ants, mice, and rats are also attracted to the smell. Most people with indoor worm bins would rather not put out the "come and get it" sign for pests by using high protein foods in the worm bin. However, some worm growers do occasionally use protein foods, like bits of chopped egg, to provide the worms and other organisms in the bin with extra nitrogen and protein.

Finally, some foods, such as citrus, with a high level of acidity can be toxic to worms and other organisms. These foods should be used sparingly, if at all. Citrus peels contain a chemical called D'limonene, which is one ingredient used in pet shampoos to kill fleas and ticks. If you would like to try feeding citrus peels to your worms, run the rinds through the blender to make a slurry. As more surface area of the material is exposed, bacteria are able to break it down faster and it spends less time in the bin. Always be certain to bury the citrus down in the bedding and keep several thicknesses of wet newspaper sheets on top of the bedding to insulate it. It also should be noted that pesticide residues on plants can be toxic to worms, so all food entering the worm bin should first be washed.

Grit

Worms grind up their food in their gizzards. Just like birds, worms need grit in their diets to help the gizzard grind the food into smaller particles. In nature, when a worm feeds, bits of soil are taken in with the organic material and the gizzard can use these bits of soil to help the digestion process. However, in worm bins that use primarily kitchen wastes, it becomes necessary to add grit. It's a good idea to sprinkle some fine soil, rock dust, or eggshells over your worm bin every few weeks to make sure there is enough grit for the worms. Some worm growers even use fine sand (but not beach sand). Beach sand can be very coarse and may contain high levels of salt.

Pathogens

Whenever wastes are recycled, safety questions arise: Can pathogens such as E. coli or salmonella be found in my worm bin, and if so, are they a problem? The latest scientific answer is that, even though these and other bacteria can live quite nicely in a worm bin, they are rarely if ever a problem—and scientists don't yet know the reason why. What follows is the current state of research.

If you are using a vegetable waste system, then most pathogens should not be a problem. Bacteria such as E. coli are generally associated with undercooked meat, and salmonella bacteria is introduced by contaminated fecal matter.

There is ongoing research concerning pathogens in industrial wastes, and so far the preliminary results are excellent. Initial research is pointing to the capability of the worm's digestive system to destroy the pathogens it ingests. So far, worms have been shown to destroy salmonella, E. coli, and other fecal pathogens in contaminated food to levels below health department guidelines. One county in Florida added worms to its sewer sludge at a waste treatment plant and found pathogens were indeed reduced to meet Environmental Protection Agency (EPA) pathogen standards. Again, these are preliminary findings, and more research needs to be done to figure out just what is going on inside the worm's digestive tract to kill the pathogens.

The EPA recommends composting possibly contaminated material at high temperatures—170°F (55°C)—for three days to greatly reduce the number of pathogens. Obviously, allowing a worm bin to reach this temperature would kill your worms, but pre-composting the materials you suspect might contain pathogens in an outside bin or pile for the recommended time and temperature before using them in your bin would be a good idea.

As far as catching anything from your worm bin, good hygiene should always be practiced when handling any soil product. Washing your hands and tools is always a good idea. Tell children who have worm composting systems in their classrooms to wash their hands after working in the bin, and for very small children, tell them not to eat it!

Pet Feces

Pet feces of any kind should never be placed in the worm bin. As we just discussed in the previous section, a worm bin cannot reach temperatures high enough to kill pathogens. Also, fresh pet wastes come under the category of green manure and can heat up your worm bin quickly, hurting your worms.

Cats can harbor the protozoa for a disease called toxoplasmosis. The protozoa are passed in infected cats' feces, and when humans come in contact with the feces they have a chance of catching this disease. Pregnant women are usually warned about this problem by their doctor and are advised to let someone else clean the litter box if they have cats. Fetuses that come in contact with this protozoa before birth can be born with serious brain damage.

Unfortunately, if you put cat feces in a worm bin, the toxoplasmosis protozoa will pass intact through the earthworm. Then, when you are cleaning or changing bedding, the protozoa can accidentally be inhaled or enter through a break in the skin. So do keep pet feces out—and don't let your cat use your worm bin as a litter box.

HOW TO FEED YOUR WORMS

• **Saving the food wastes.** First, you must collect the kitchen wastes to feed your worms. Many people put a plastic bucket under the sink or keep an extra small trash can next to their regular trash container. One good idea for a container is to recycle a one-gallon plastic milk jug. Cut the top off so it's wide enough to get food wastes in easily, but leave the handle intact. When this container starts to get grimy, toss it out in the recycling bin and cut another one. Actually, any container is fine as long as you don't use a lid on it. Placing a tight lid on the container keeps fresh air out, thereby promoting the growth of anaerobic bacteria.

Anaerobic bacteria grow in environments that lack oxygen. These bacteria do a far less complete job of breaking down nutrients than aerobic bacteria and usually leave lactic acid or ethyl alcohol as their end products. These end products increase acidity, which can be harmful to worms, so it is important to try to limit the growth of anaerobic bacteria in your bin.

If you are bothered by flies or other insects buzzing around your waste container, make a fine screen lid. Use a piece of screening from an old screen door or window and tape it to a piece of thick cardboard that has a hole cut out of the center. Of course you can get fancier with your lid project, but this will work.

• **Preparing the food.** Most kitchen wastes can be placed in the bin without any preparation. However, the outer leaves of some vegetables have been sprayed with pesticides; these should be washed before they are fed to the worms. Washing everything is a good habit to get into, so you don't have a toxic problem later. Some of the vegetables and fruit that are known to have pesticide residues are strawberries, cucumbers, peppers, cabbage, and apples.

Many worm growers recommend chopping or grinding food in a food processor or blender to make the food easier for the worms to eat. This is up to you: Larger pieces need to decompose a bit before the worms can eat them and will thus take longer to recycle—but they will be recycled.

Worm food or feed stock needs to be moist before going into the worm bin. This is generally not a problem with vegetable wastes—if anything, vegetable wastes can sometimes be too wet. If you have dry kitchen waste, such as dry bread, soak it and wring it out to a damp consistency. You can mix dry food with vegetable foods that have a large moisture content to produce a moist diet. Some growers who feed a dry commercial food will sprinkle the food in the bin and then sprinkle water over it. This is easier for them.

One word here about water: A number of worm growers caution against using chlorinated water in your worm bins. The chlorine, they say, will eventually harm your worms. If you live in an area with chlorinated water, you can buy dechlorination drops at any fish store or use distilled water. However, if you are feeding fairly wet vegetable food to your worms, you may not need to add additional water.

• **Where to put the food in the bin.** If you have purchased a commercial worm bin, then follow the directions given with the bin. One commercial worm bin has you spread the food evenly over the bottom worm tray. If you have constructed your own worm bin or built one of the bins described in this book, then you should feed your worms using an area method.

Divide your box in sections or areas of feeding. For small boxes, these areas can be very simple. Try numbering each corner and start feeding first in area one, then area two, and so on, until you are back to one again. If you are feeding once a week, it will take you four weeks before you get back to one again. Hopefully, by the time you get back to the first area, the food will have been eaten by the worms and will no longer be recognizable. It's a good idea to make a note of where you fed the worms last, so you don't put the food in the same area again. A simple record sheet or piece of colored tape that you can move from corner to corner will work. Record sheets can also include information on what kind of food and how much was fed to the worms at each feeding. This can help you find out what kinds of foods your worms like and how much is enough to feed them. Larger bins have more areas in which to place the worm food. Make a map of your feeding areas and keep it by the bin. One way to do this is to divide the surface area of your bin like a tic-tac-toe board. Slip the map into a plastic sleeve and use an erasable marker, grease pencil, or reusable sticker to mark the area last used. When it's time to feed the worms again, move the sticker or re-mark the map. These sleeves can be mounted to the side of the box or on the lid for convenience.

• **How often should you feed the worms?** There are no special hard-and-fast rules about when you should feed your worms. Some people like feeding them twice a week because they don't like the smell of their kitchen waste container if they don't empty it often. Others will wait and feed the worms once a week. It really doesn't matter when you feed the worms as long as they are getting enough food. If you're not sure, check the bin daily at first and watch the food gradually disappear. This is an especially good idea when baby worms begin to hatch and feed, to make sure they are getting enough food. When the food is almost gone, add food to the next area. As you gain experience caring for the worms, you will know approximately how much food they need.

If you are worried that opening your bin to look at the worms is disturbing them too much, remember that they can't see red light. So, if you check your worms at night using a flashlight with red cellophane over it, they will never know you have been there.

• **How to put the food in the bin.** When you have determined where to put the food, pull back the top layer of bedding using a small hand trowel and bury the food about two inches deep. Replace the top bedding. Covering the food controls odors and makes the bin less attractive for flies and other pests. Some worm growers prefer to scatter the food on the top of the bedding and then add more bedding on top. Experiment and determine what works best for you.

A note about adding fruit juice to your bin: Sometimes when fruit juice is added to a bin and dry bedding is placed on top, the juice will wick its way up. You end up with juice-soaked top bedding and every fruit fly in town coming to your house for dinner. If you have leftover juice to feed to worms, go ahead and pour it in, but place a couple of sheets of wet newspaper on top of the juice. The juice is less likely to wick through wet paper. Replace your bedding on top of the newspapers.

• **How much food waste can a worm eat?** As we calculated in chapter 3, generally speaking, two pounds of earthworms will eat one pound of organic material or kitchen wastes in twenty-four hours. Under perfect conditions, this amount could rise to two pounds of earthworms eating two pounds of kitchen wastes; and under poor conditions the amount will fall to less then one pound in twenty-four hours. Watching the amount of food consumed by the worms is one way to tell how well they are doing.

TEMPERATURE, MOISTURE, pH, AND AERATION REQUIREMENTS

The environmental conditions in worm bins are quite important. In nature, if the soil gets too hot or cold, the worms can regulate their body temperature by moving down deeper in the soil. If the soil gets too acidic, the worms can simply move to another area. In keeping a worm bin, we are responsible for making sure our worms have tolerable living conditions. Drastic changes can send worms into shock and they can stop eating or breeding. Worms that have gone into shock may take some time to recover, and some may never recover and will die. Monitoring your worm bin is essential in keeping track of living conditions. You'll know when a problem is occurring, so you can take corrective measures.

• **Temperature** Worm bin bedding should be kept at temperatures from 55 to 77°F (15 to 25°C). The optimum temperature for worms is between 72 and 74°F (around 23 to 24°C). Use a thermometer to check temperatures in the winter and summer. Special compost thermometers are also available—these have stems that are two feet long. Temperatures over 80°F or under 55°F in the worms' environment will greatly slow down their activity. Redworms will be severely stressed when temperatures fall under 50°F (10°C). If you live in an area where it becomes very cold in the winter, consider moving outdoor worm bins inside or adding extra insulation to the bin for the winter.

Freezing temperatures can kill your worms.

Bedding temperatures above 86°F (30°C) can also be harmful to your worms, and anything above 90°F can be lethal. If your bedding is starting to get too hot, check the location of your worm bin. Is it in the sun or in a poorly ventilated area, or are the air holes in the bin blocked? Consider moving your bin to a basement or back closet of your house. Add extra bedding and keep air flowing in the system. The moisture in the bedding will evaporate with good airflow and keep the temperatures down. Watch moisture levels carefully and keep the levels up with extra water if needed.

• **Moisture levels** Moisture levels are very important to worms. Their bodies contain approximately 75 to 90 percent water. Worms breathe by taking in dissolved oxygen through their skin, which must be kept moist. If worms dry out, they can no longer breathe, and they will die. Studies have shown that worms can live completely submerged for over an hour in oxygen-rich water, but, as soon as the oxygen is removed, they die.

Moisture levels in bins should be kept between 70 and 80 percent. This is the optimum moisture for worms. In very dry climates, bins may be a bit drier. Try not to let the moisture level dip below 60 percent. In summer months, consider placing a piece of plastic over the top of the bedding to help keep moisture in. If you use plastic, make sure there are enough air holes in the sides of your bin.

In an outside worm bed or windrow, worm growers report a good top moisture level is between 35 and 45 percent moisture. This ensures adequate fresh air in the soil. However, some worm growers recommend an average moisture level of 55 percent.

Check your worms at different moisture levels and chart their activity: Are they eating well, reproducing? Decide what moisture content your worms like best. Moisture meters or a moisture probe can be purchased at nurseries or electronic stores to check moisture levels and are a good investment for the worm grower.

• **pH levels** pH is a chemistry term for "potential hydrogen." pH measures acidity to alkalinity, on a forteen-point scale. On the pH scale, the number one indicates the most acidic, the number fourteen the most alkaline, and the number seven is neutral.

So, if you measured a sample of soil to be six and a half on the pH scale, you would say it is slightly acidic. If you have a swimming pool, most likely you have checked its pH levels yourself or seen someone else do it.

Most plants and animals do well in the pH range of six to eight. This is also true of worms. If you could keep your worm bin at a pH of seven (neutral), you would have happy worms. However, depending on living conditions and which foods your have fed the worms, the pH will fluctuate. A pH meter can be purchased to check the pH of your worm bin easily. Chemical test kits are also available, but these are messier, and one mistake voids the whole test.

• **Aeration** In chapter 3, we talk about the breakdown of organic material and how this takes place in a composting system. In a worm bin, we strive to keep oxygen levels high to keep the aerobic bacteria happy and eating the organic material. Oxygen is also necessary for the survival of our worms. For a worm to breathe, oxygen must dissolve in the mucus on the worm's skin. Without oxygen, our worms die and the aerobic bacteria die, setting the stage for a takeover of the worm bin by anaerobic bacteria. A good indication of anaerobic activity is the presence of odors. Keeping bedding loose and providing air and drainage holes in the bin are all good ways to ensure enough oxygen is getting into the bin. Some worm growers will actually turn their beds over every now and then to get added oxygen deep into the beds if they feel the bedding is getting compacted.

HARVESTING

Finally, you have the worm bin set up and going. Your worms are eating your organic wastes and now there is little or no original bedding visible in your bin. The original bedding has changed to brown and earthy-looking vermicompost. Vermicompost is a mixture of worm castings and decomposed organic matter. This process can take anywhere from six weeks to four months, depending on the size of the bin and the number of worms in it. The volume of the bedding will decrease over this time, making the environment more and more hostile for the worms. It's time to harvest. What a great day! Now what do you do?

Worm growers who have purchased commercial worm bins usually don't have to worry about harvesting too much. These worm bins come with specific instructions, which work quite well, about when and how to harvest the worms and castings. For those who have made their own box-type bin system from either plastic or wooden boxes, there are several different ways to harvest the castings and the worms.

NOTE: When harvesting worms and vermicompost, some harvesting methods involve moving the worms from one bin to another. If you are planning to move your worms, you should always have the new bin bedded and ready to receive the sorted worms before you start harvesting, to reduce the stress on the worms.

• **Sorting method**: Many worm growers who have small bins feel separating the vermicompost and worms in the bin is an easy method. One benefit of this method is that it doesn't disturb the worms too much, and it allows you to remove as little or as much of the vermicompost as you want.

Begin by carefully separating the vermicompost from the uneaten food and the worms. Use a paintbrush to move the vermicompost to one side. Scoop the compost out. Wait a few minutes for the worms to move deeper into the bin and repeat the process until all the vermicompost is removed. The last layer will be thick with worms. If you need to remove some worms, do so now, or place new bedding on top and start composting again.

• **Screening method**: This method involves separating the worms from the vermicompost using a screen. One way to do this is to dump your worm bin's contents, a bit at a time, on a framed, $1/4$- or $3/16$-inch screen. The vermicompost will drop through, leaving worms and larger particles of food and bedding behind. Return the worms back to a clean and freshly bedded bin or bins. Keep in mind that this method does stress the worms, and it may take them a few days to recover.

Another method involves using a screen inside the bin. This method works great for small bins. Place a piece of flexible screen ($3/16$-inch or so) on top of your worm bin when it is time to harvest. Use a big enough piece of screen so the screen will go up the sides of your bin. You will use this extra screening to lift the top section out later. Now, rebed the worm bin right on top of the screen. This now becomes the top part of the bin. The worms will now move up through the screen to feed on the new food and bedding. When the worms are established in the top bedding, lift the top part out of the bin by the screen. Dump out the fine vermicompost in the bottom and pour the top part back into the bin. Repeat this process again when the bottom part of the bin is ready to harvest.

• **Light method**: Worms are photophobic, which means they do not like light. To use light to help you, dump your worm bin in a tall pile on a piece of plastic. (Larger bins can be dumped a bit at a time.) Shine a light directly overhead—or do this outside on a sunny day—and the worms will burrow back down into the pile to get away from the light. Now brush off the vermicompost or castings off the top and sides of the pile and place them to one side. When the worms are exposed, wait until they burrow back down into the pile. Do this until the worms are completely separated from the compost. Now you have a pile of worms to separate into new bins or sell. Be sure to have new bins ready to receive the worms right after sorting, because worms exposed to light for too long will die.

• **Moving method**: This method, which works well for larger bins that are too heavy to lift, involves moving the worms in your bin from one area to another. When the bin is ready to be harvested, move the contents of the bin over to one side. Now rebed the empty side with the same kind of bedding you used initially. Start feeding the worms on the new side only. The worms will start moving to the new side for food and clean bedding. This will take several days. When the worms have moved over, harvest the old side. Add new bedding to the old side and repeat when necessary.

•**Water method**: Some worm growers prefer to use their worm castings in liquid form. One way to do this is to harvest the castings by dissolving them in water. Spray your worm bin with water and collect the casting water that comes out, being careful not to waterlog your worms for too long. Rebed the worms as quickly as possible.

Another way this can be accomplished is to use the moving method described earlier. When the contents of the bin are moved over, add a screen partition to the bin. Now rebed the other side and feed the worms there. When the worms have moved over, slide a solid piece of wood next to the screen partition. This partition should be tight fitting and should not allow much water to pass between the two sides. Tilt the bin and spray water on the finished compost side. Be sure to have a drain outlet on each side of the worm bin. The collected liquid fertilizer can then be poured through a fine strainer to retrieve any worms or egg capsules.

• **Death method:** Some worm growers do not want to sort, separate, or do anything but get high-quality vermicompost that is almost completely vermicast or castings. Vermicast is material that has been worked and re-worked until it has very little decomposed organic matter left in it. It is not considered to have more nutrients than vermicompost, since the material has been worked so much by the worms, but it has a very smooth texture that is quite desirable. To use this method, set up the worm bin and feed it over a period of time. Get the worms off to a good start for a couple of months, and then don't feed them for three or four months. By that time, most of the worms will be dead and only castings will remain. Worm growers who prefer this method must buy new worms every time they set up a bin.

• **Modified death method:** This method improves on the death method by allowing some worms to live. To use this method, you will need two worm bins. Start one bin, and when it is almost time to harvest it, place some food that is very attractive to the worms, like a cantaloupe rind, on top of the bin. Draw as many of the worms as you can to the food. Now dig down and transfer these worms to the new bin and start feeding them. The old bin, which will still have some worms left in it, is now not fed, and the worms are allowed to work and re-work the bin until they eventually die. This process will take a bit longer than the death method because there aren't as many worms working in the bin that is not being fed. The benefit is that you still have a working bin eating your wastes and another bin producing well-processed castings.

• **The garden method**: If you are raising worms for use in your garden, then this method couldn't be easier. When your bin is ready to be harvested, simply remove one-half to two-thirds of your bin and take it to the garden. Your garden needs will dictate where you put the worms and compost. Make sure there is enough organic material for your worms to survive. Now rebed the rest of the worm bin and continue.

Worm egg capsules or cocoons

A word here about worm egg capsules. They are very small and will easily fit through a 1/4-inch screen. When you separate the vermicompost from the worms, it is very hard and time-consuming to pick out the egg capsules from the compost. One way to get around this is to place the separated vermicompost that contains capsules in another bin. In the center of the bin, place some blended kitchen wastes. Wait about two months, checking it from time to time, and then go back and screen it again. By then the baby worms will have hatched and can be easily separated.

There are several references that say you should not touch egg capsules with your bare hands. One states, "oils from touching the capsules may cause mold and infertility," and another says, "natural skin oils will break the surface seal of the egg capsule." So, it is a good idea to wear gloves if you plan to hand-sort worms. (Besides, some people do not like the slimy feel of the worms themselves!)

WHAT SHOULD YOU DO WITH EXTRA WORMS?

This is a wonderful problem to have because it means you are definitely doing something right with your worm bin. If you only want to maintain one worm bin for home waste disposal, and you live in an apartment or don't have a garden, then the problem of extra worms may come into play.

HERE IS A LIST OF JUST A FEW OF THE PLACES AND PEOPLE WHO WOULD WANT YOUR WORMS:

Gardeners

Garden centers, nurseries

Dairies and horse ranches
 (to vermicompost their manures)

Fishermen, bait shops

Fish hatcheries

Game bird farms

Pet stores

Scientific researchers
 (who need them for experiments*)

Commercial worm growers

People wanting to start their
 own worm bin

Schools, 4-H clubs, Boy and Girl Scouts

Waste treatment plants, landfills

*Experiments are underway using earthworms to dispose of the vast quantities of garbage and sewage sludge produced by cities and turn it into safe and useful fertilizer.

Worms, which are high in protein and low in fat, are being studied as a future food source for developing countries. Check chapter 13 for recipes featuring earthworms.

5

PROBLEMS IN THE WORM BIN

New worm growers think that if they follow all the instructions that come with their new bin, they won't have any problems with their worms. Well, new worm growers will soon learn that growing worms requires special techniques, and problems sometimes arise. It's the ability to recognize these problems and act quickly that separates the successful worm grower and the not-so-successful worm grower.

Every worm bin or bed is unique, so your bin will probably act a bit differently than everyone else's. Factors that will be unique to you might be, for example, the temperature and humidity of your local climate and your worm bin's location. Factors that worm growers have in common may be moisture levels in the bins, types of foods being given, and the type of worm bin being used to grow the worms.

It is estimated that only 7 percent of the people who own worm bins will experience no problems with their bins.

The following is a list of common problems that worm growers face and practical solutions that worm growers can use:

AN UNPLEASANT ODOR IS COMING
FROM MY WORM BIN.

Odors can be caused by several factors:

1. Your worm bin is over-full with food wastes. The worms simply have too much food to eat and the leftover food is rotting, causing the odor. To correct, break up any clumps of food and stop adding food until the worms have had a chance to eat what you have already provided. You also should try covering the food well with bedding material or sheets of damp newspaper if you are not already doing so.

2. Your worm bin isn't getting enough air. Gently stir up the entire bin (if your bin isn't too large) to get more oxygen into the system. Anaerobic bacteria, which grow in the absence of oxygen, are usually quite smelly. You may need to repeat the stirring every so often for several days. Also try mixing in some fresh loose bedding to get more air into the system. Check to see that you have enough air holes in your bin and that they are not blocked.

3. Your bin is too wet. Check the drainage holes to see if they are blocked. Blocked air holes can also cause moisture levels to rise and the oxygen levels to fall. If the drainage holes are not blocked, you might be adding too much wet food to the bin. Each time you put food in the bin, place a layer of damp shredded newspaper over the food. This helps to absorb excess moisture in wet food. Also, avoid adding any extra water or foods with a high moisture level for a while. (Foods such as puréed fruit are very wet; use dryer foods instead.) Adding some dry bedding could also help. If your bin is really wet and the worms are in danger of drowning, soak up as much water as you can with a turkey baster and push sphagnum moss or paper toweling down the sides of the bin to absorb more. Change the moss and paper often until the moisture level comes down.

4. Your bin is too acidic. Acidic conditions may cause an odor. Test your bin with a pH meter available at many local nurseries, home centers, and swimming pool supply centers. Cut down on the amount of acidic foods you give the worms. Try adding smashed eggshells to lower the pH. (Adding eggshells a couple of times a month will go a long way toward balancing the pH.) Other safe products that can neutralize acidic bedding are rock dust and calcium carbonate. Both of these products will also provide grit to the worms. Some people like to use garden lime to lower pH, but you should be very careful, as it can lower the pH dramatically if you use too much. Hydrated lime should never be used because it can severely shock the worms.

5. Some foods just stink. Foods in the broccoli family just plain stink and you may wish to avoid using them. Try one food at a time, and, if the smell bothers you, don't use that food again.

6. You are using the wrong food. Meat products, dairy foods, and very oily foods should not be fed to worms. They usually go rancid while they are decomposing and cause a terrible odor.

THE WORMS ARE LEAVING THE BIN.

Worms used for worm composting systems prefer specific conditions and some worm species are choosier than others.

1. Do you know what kind of worms you have? When you order worms, be sure to ask for the scientific name because many worms have several common names depending on whom you ask. Find out for sure by asking for the scientific name and then looking up the specific conditions for that worm species in chapter 4 of this book. Now that you know which worm you have, check the moisture levels, temperature, and acidic conditions specific to your worms.

2. Use light to keep them in the bin. When worms are shipped or sorted, they become stressed. When these

stressed worms are placed in the new bins, they will sometimes stage a walkout for no apparent reason. Conditions could be right, but they might leave just the same. Leave a bright light on over the new bin for a few days to force the worms into the bedding until they get used to their new home.

THE WORMS ARE DYING.

You notice the number of worms seems to be dropping, and you have found some dead worms. This is an emergency situation, so act quickly.

1. Move your worms to another bin now! It is always good to have an emergency bin ready in case of emergencies like this, but if you don't have one, don't panic. Use anything on hand and don't worry about the size: an old ice chest, wooden box, or garbage can can work in a pinch as long as it is clean. Fill the bin with a neutral bedding like clean leached peat or shredded newspaper. If your worms were once happy in a specific bedding, use that. Once you have the worms moved to the new bin, check your bin for these possible problems:

    a. The bin is too wet and the worms are drowning. Soak up as much water as you can and insert small rolls of newspaper throughout the bin.

    b. The bin is too dry. Give the bin a good soaking in dechlorinated water. Make sure the drain holes are open.

    c. The worms are not getting enough food. When worms eat everything there is to eat in the bin, they will begin to eat their own castings. This can be fatal to the worms; in a case like this you should harvest the worms as soon as possible.

    d. The temperature is wrong in the bin. Worms usually like temperatures from 55 to 77°F. Use a probe thermometer to determine the internal temperature, and move the bin if necessary.

    e. Check light levels. Too much light can kill worms or make them uncomfortable. A nice dark basement is an excellent place to keep your worm bin.

f. Are you using chlorinated water? Some communities have a chlorinated drinking water system; this can sometimes be harmful to worms. Switch to unchlorinated water.

g. The pH is wrong. The bin can be too acidic or too alkaline. Check with a pH probe. A pH of 7 is ideal.

Finally, if you can't find anything wrong with your bin, it might be easier just to dump out the old bedding and food and start again. This time, monitor the bin closely for any changes. If you are not doing so already, try keeping records to pinpoint the problem.

HOW DO I GET RID OF THE FLIES AROUND MY BIN?

When most people complain about flies, they are talking about small fruit flies (*Drosophila*). These we will address in more detail in chapter 8, but, to keep them away from your bin, try these helpful suggestions.

1. Put a lid on your bin. If you can't do that, then lay newspaper, old carpeting, plastic sheeting, or a fine screen on top of the bin. (Make sure the bin gets plenty of air if you use plastic sheeting.)

2. Make a trap for the flies. A small bowl half-full of apple cider vinegar mixed with a drop of liquid dishwashing detergent makes a good trap. Drawn to the smell of the vinegar, the flies fall in the liquid and drown.

3. Bury the food. Just like their name implies, fruit flies love the smell of rotting fruit. Bury the food completely so the flies won't be so attracted to your bin.

4. Vacuum the bin. This may sound strange, but a small hand-held vacuum will eliminate the flies with one pass.

WHY IS IT TAKING SO LONG FOR THE WORMS TO PRODUCE ENOUGH CASTINGS TO USE IN MY GARDEN?

Most people don't recognize that worm bins are mini-ecosystems in which everything must go along at its own pace. You don't just get a new worm bin and throw in bedding, worms, and food and have the worms eating all of your household vegetable wastes and producing enough usable castings in one day. But you will—just give it time! Most worm bins need to be harvested every two to four months, depending on the size and type of the bin. Worms need attention and time, and if you set things up right you will have great castings soon enough.

THE WORMS AREN'T EATING.

Worms definitely like conditions to be just right, but some things to consider are:

1. *Is it a new bin system?* Sometimes when you first put worms in a new system they are slow to eat. They need to get used to their new home. It is best just to wait a while and avoid feeding any more food (or you take the risk of adding too much food). The worms should start eating in a few days.

2. *Check conditions.* Go through your checklist of requirements for the species of worm you have. Are the temperature, acidity, and moisture levels within limits? Is the bin getting enough air? Check everything.

THE BIN IS TOO ALKALINE. WHAT CAN BE DONE?

1. *Move the worms to a neutral bin.* If the bin is very alkaline, that would be the easiest and quickest way to correct the problem. Then start your main bin over with fresh bedding.

2. *Feed an acidic food.* When the bin is only slightly alkaline, feeding the worms an acidic food for a while will help bring down the bin's pH level. Try feeding the worms some citrus (without the peels), but again, not too much at a time. Test the bin's pH level regularly until it gets back to normal.

3. *Add fresh neutral bedding to the bin.* Adding some shredded paper or leached peat moss to the bin will also help lower the pH level.

4. *Add sulfur.* Worms like very gradual changes in their bins. This might be tried as a last resort and then used very sparingly.

THE WORMS ARE GONE.

This is a horrible discovery. Many things could have happened in an enclosed indoor worm bin.

1. *The worms died.* Yes, this could happen. Since worms decompose very quickly, you might miss seeing this if you weren't watching your worms closely.

2. *They are hiding.* Sometimes when conditions are right you will find the worms all bunched up together in a corner or bottom of the bin. Look carefully again to make sure they are really gone.

If your bin is kept outdoors or your worms are in an outside bed, then other possibilities come into play.

1. *The bed was too hot.* If you put grass clippings or other green waste into your worm bed, the temperature of the composting waste may have raised the temperature of the bed and forced the worms to leave. Most of the time, when the temperature goes down the worms will return.

2. *They escaped.* If your bin is on the ground, then the worms may have simply left the bin through the drainage holes. This has happened to more than one new worm grower.

LARGE GRUBS ARE IN MY BIN.
HOW CAN I GET RID OF THEM EASILY?

Many people who use grass clippings as a food may run into this problem at one time or another. The grubs are not hurting the worms, but instead are competing for the food in the bin. In an outdoor bed, many people don't mind the grubs, especially if they are trying to compost large quantities of green wastes. But if you don't want them, try placing a melon rind, skin side up, in your bin. The grubs love this food and will come up to feed on it. Then you can simply remove the rind, along with the grubs. Melon rinds will help draw worms up also.

HOW CAN I TELL IF MY WORMS ARE HEALTHY?

New worm growers ask this question all the time. A few ways are:

1. *Get to know your worms.* As any good zookeeper or rancher will tell you, you need to get to know your animals. By observing healthy worms, you will soon be able to spot unhealthy ones.

2. *Shine a light on the worms.* When you open your worm bin, shine a flashlight inside. The worms should quickly bury themselves. If they don't, they could be sick.

3. *Check slime levels.* All worms have slime on their bodies. If your worms are looking dry, then you have a problem.

4. *Are the worms eating their normal amounts?* If they are not, then the worms may be having a problem. Check conditions in the bin.

5. *Smell.* Foul odors can indicate a problem.

MUSHROOMS CAME UP IN MY BIN TODAY. IS THIS BAD?

Mushrooms strongly indicate that fungi are present; they will do no harm to your worm bin. Fungi are part of the decomposition web and, since they are microscopic, most people do not realize they are even there until a mushroom comes up. Mushrooms have one bonus: their appearance is a good indicator that the temperature of your bin is between 70 and 75°F, which is a great temperature for redworms. Of course, it should go without saying, you should not eat these mushrooms. Simply pick them out and throw them away.

MOLD IS GROWING ON FOOD I PUT IN THE WORM BIN. IS THIS OK?

Mold is another organism in the composting process. Most people who bury their worm food do not see the mold doing its part for decomposition. The mold becomes noticeable when you don't bury the food. Usually this is no big deal and nothing to worry about, unless you are sensitive to mold spores. If you are allergic to molds, then vermicomposting indoors may not be for you. Keeping your bin outdoors and burying the food well will help.

6

Other Animals Found in a Worm Bin,

or

Worm Bin Visitors: Good Guys and Bad Guys

ANTS

You go to tend your worms as usual and what should you see but a stream of ants going into your prized worm bin. Your first thought is to grab the ant spray and spray the life out of the creatures. Don't! Insecticide sprays are toxic and dangerous to your worms.

Ant eradication and prevention ideas

Ants are always searching for food or a home, and your worm bin looks like a good place for both. They probably won't do any real harm to your worms, but they are a great nuisance. To get rid of the ants you see, spray a commercial window cleaner containing ammonia or lemon extract on a paper towel or rag and wipe the ants off the bin. Wipe only the outside, so the cleaner does not get into the bin. The cleaner will break the scent trail of the ants. Spray the cleaner on the trail away from the bin to kill the ants and remove the scent trail. When the cleaner dries, you can easily sweep or vacuum up the dead ants.

Follow the trail to find out how the ants are getting into your house and to your bin. Are they coming in through a hole or a crack in a window, or just walking in the back door? Prevention is the key. Plug any holes or cracks and repair screens. Discourage ants from coming in by using a barrier. There are many substances ants will not cross, and you can use one of these to block them from entering:

- cayenne pepper
- cleanser
- eucalyptus oil
- lemon juice
- lemon-scented oil
- mentholated rub
- talcum powder
- tanglefoot or any sticky substance
- water

Experiment with these; maybe you know of some others. Unfortunately, they will not last forever and must be reapplied periodically. Bins that are on legs can be made ant-proof by placing the legs in containers of water, like coffee cans or cut-down plastic milk jugs. Place water about one-half to three-quarters full. Keep the containers and water clean of debris. You don't want a bridge for the ants to cross.

Control ants by sending poison bait back to their nest. The ants will find the bait and return with it to the nest. There they will feed the bait to the rest of the ants, killing all of them.

A simple bait that can be used for ants inside and out is:

ANT DEATH BAIT

1/4 cup granulated sugar
1/4 cup boric acid or borax
a handful of dry dog or cat kibble
old spice jars with shaker tops,
    or small plastic containers with lids

Mix the sugar and boric acid or borax together. Clean and dry an old spice jar. If you don't have an old spice jar, use a small, clean plastic tub, like the ones margarine comes in. If you are using the plastic tub, poke at least four holes: one in each direction, about 1/4 inch up from the bottom of the container. Label containers "Ant Bait."

Place about 1 heaping tablespoon of the boric acid/sugar mixture in the jar or tub. Add 2 or 3 pieces of the kibble. With smaller kibble use more. Snap on the plastic shaker top or the lid of the plastic tub. Recipe makes several ant bait stations, or store unused bait in a sealed container for later use. Make sure to carefully label all containers "Ant Bait"!

Place the jars or tubs where ants will find them (turn the spice jars on their sides so that ants can easily crawl in). Keep them away from pets and children.

Some ants like liquid baits better then dry baits. To prepare a liquid bait, mix nine parts corn syrup with one part boric acid or borax. To use this liquid bait, prepare the plastic container as in the recipe above, making the holes in the container high enough so the syrup does not pour out. Put a quarter-sized dollop of the bait in the center of the container, and place it where you see ants. When the bait dries out, replace it with fresh bait.

If you can, follow the ant trail back to the nest. Here, you have several options to kill the nest, all of which are nontoxic to the environment.

1. Pour boiling water on the nest. This won't kill all of the little creatures at one time, but it will certainly put a dent in their population.

2. A strong hot chile solution will not only kill the ants, but will make the nest unlivable.

3. Sprinkle dry cornmeal around the entrance to the ant nest. This old remedy is said to work by expanding in the ant after it is eaten, killing the ant. (It's the same reason why we don't throw rice at weddings anymore. The rice will do the same thing to birds.)

HOT PEPPER SOLUTION

Suntea jar or any large jar, at least 1-quart size
2 to 4 sliced hot peppers
(serrano, habanero, or the hottest available in your area)
1 quart water

Put the sliced peppers in the jar and fill it with hot water. Let the mixture steep for at least 24 hours. Remove the peppers and pour the solution into the nest.

Finally, ants can get into your worm bin if you use fresh garden wastes in your bin. These little creatures will come right in with that handful of grass clippings. If you do use garden wastes that are not composted first in your indoor bin, it is best to kill the ants. To do this, either soak or heat the greens prior to adding them to the bin. Soaking works great for leaf material, but it's hard for grass. One way to heat grass easily is to place it in an old aluminum roasting pan (the level of the grass should-n't be higher than the top of the pan) and pop it into a warm oven. Thirty minutes in a 180 to 200°F oven should do the trick.

A word about ants

Ants are members of the wasp order of insects called Hymenoptera. There are more than 3,500 different species of ants that live in nearly every land habitat in the world. All ants are social in nature and live in colonies called nests or mounds. Colonies consists of a queen (or queens) and female workers. Males are only produced by the queen when the nest is getting crowded. Then she will produce winged males and females that go out and establish new colonies. Female worker ants have many jobs in the nest. Some workers tend the queen and others tend the young; still others are scouts.

Scout ants are workers that forage for food. When a scout ant finds food, such as in your worm bin, she will lay down a scent trail that the rest of the worker ants can follow back. These scent trails only last for a few minutes, but that is usually long enough for the ants to get from the nest to the food.

Ants are the geniuses of the insect world. Scientists have shown that ants are capable of individual learning and passing on what they have learned to other ants. They can display memory, correct their mistakes, and communicate among themselves. Ants communicate by exchanging chemicals in their mouths when they meet.

CENTIPEDES

While tending your worm bin, chances are you will come into contact with this prehistoric-looking creature. Centipedes are fast moving and they are looking for a meal. Unfortunately, it's your worms they are hoping to eat. It's like having a wolf in the hen house.

What are they?

Centipedes are arthropods and members of the class Chilopoda, which has 3,000 described species. They can be found throughout the world in both the temperate and tropical regions. They live in soil and humus and can be found under stones, bark, and logs. Centipedes can range from one inch to one foot in length and come in various colors. They can be recognized by their flattened shape and their single pair of long legs per segment. (The name centipede means "hundred legs.")

The whole class is believed to be predators of living animals. They use a set of poison claws to capture, stun, or kill their prey. The venom that centipedes possess is not sufficiently toxic to be lethal to humans, but a centipede's bite can be quite painful. People who have been bitten say it is like a severe yellow jacket or hornet sting.

Getting rid of centipedes

Any centipedes you find in your worm bin should be destroyed. They are definitely pests, and they are out to eat your worms. There is no magic here about getting rid of them: When you see a centipede, chase it down and kill it with a garden trowel. Some worm growers say you can pour a bit of soda water on the centipede to stun it, but they are fast runners.

One good thing is that centipedes, like most predators, are territorial. This means there won't be too many in any one spot.

ENCHYTRAEID WORMS

Enchytraeid worms, sometimes called pot worms or white worms, are relatives of the earthworm and belong to the family Enchytraeidae. They are small worms, reaching only 1/4 inch long when full grown, and are often mistaken for insect larvae or newly hatched redworms. Newly hatched redworms are transparent; within hours of their hatching they show a visible red vein running the length of their body and appear reddish. In contrast, an enchytraeid's blood contains no hemoglobin and so they remain white throughout their lives.

Enchytraeids can be found anywhere earthworms are found. One scientific study showed that in one square meter of meadow soil there were 700 earthworms and 8,000 enchytraeids.

Enchytraeid worms, like earthworms, eat decomposing plant material and burrow through the soil. Their castings become food for other microorganisms needed for decomposition. Many people feel they are quite beneficial in changing organic waste into compost, while others feel they are competing with redworms for food and space and should be eliminated in the worm bin. If you have enchytraeids in your worm bin and they are doing a great job producing vermicompost and eliminating your kitchen wastes, then don't sweat them being there.

> Scientists have calculated that if two houseflies met and mated and no predators ate them or their offspring, the fly pair and their offspring could produce enough flies to cover the entire earth 47 feet deep with flies—in just 1 year.

FLIES

 We are all quite familiar with the irritating buzzing of flies. There are thousands of different species of flies and there could be several different species hovering around your worm bin right now. Some of the more common ones are: house, vinegar, minute, and fruit flies, just to name a few. All of these are capable of making a nice home in your worm bin. What can you do?

Keep flies out!

Preventing the flies from taking up residency in the first place is your first line of defense. It is much easier to keep them out than it is to get rid of them once they are there. Some things you can do are:

1. Place a screen cover over your kitchen waste container. Keep the flies out of your worm food before you feed your worms. Do not allow flies to get to the food waste

and lay eggs in it. If you feel that flies have already gotten to your kitchen wastes, all is not lost. Place the waste in an old baking pan and cover it with foil. Heat the waste in a 200°F oven for 30 to 45 minutes, depending on how much waste you have. If you have a lot of waste, heat it for 45 minutes.

2. Bury all food. Burying the worm food in the bin, at least two inches down, helps prevent the flies from getting to the food and laying eggs in it. It also prevents attractive food odors from escaping and calling the flies to dinner.

3. Do not bury the food in the same place twice. This way you'll prevent piles of uneaten food that can attract flies.

4. Feed worms a varied diet. A balanced diet will help keep the conditions of the bed from becoming attractive to flies. Citrus is one food item that can make the bed more acidic and appealing for flies.

5. Don't overfeed. Uneaten food attracts flies.

6. Cut food into small pieces. This will help the worms eat the food faster and make it less attractive to flies.

7. Place a thick layer of newspaper, carpet, or cloth over the surface of the bin. This will keep the flies from coming in contact with the worm bedding. Wetting the newspaper will make a better seal. A piece of plastic used to keep moisture in will work, too. In outdoor windrows, a layer of dry leaves or grass can keep flies from coming in contact with the bed.

8. Keep the bedding material from becoming too wet. Wet conditions encourage fruit flies and anaerobic bacteria.

9. Never place food that has maggots in it in your worm bin. (This may sound silly to mention, but it has happened.)

Getting rid of the flies in the bin

Once flies have established themselves in your bin, getting rid of them is much tougher. Many species of fly and gnat spend the first part of their lives living in compost as maggots. In this form, they are doing just what your worms are doing: eating organic matter. In nature many animals live and eat together in the soil, but, in your worm bin, maggots are competing with your worms for food. So, it's time to take action. What can you do besides changing all the bedding in your bin?

1. Check conditions in the bin. Are they too wet or acidic? If they are, take corrective actions. See chapter 7.

2. Remove rotting food from the bin and as many of the maggots feeding on it as possible.

3. Stop feeding the worms for several weeks to allow the fly maggots to pupate, become adults, and fly away.

4. Bury food deeper in the bin. The food may have been buried too close to the surface of the bin where the flies could find it. Flies cannot dig, but, if they detect an odor of food, they will lay their eggs on the bedding.

5. Make a trap for the adults. There are many commercially available fly traps—or you can make one yourself. A few examples are:

   FLY TRAP #1. A solution of one cup apple cider vinegar with a drop or two of liquid dish washing detergent mixed into it makes an excellent bait for fruit flies and can be added to several different traps.

   a. Pour it into the bottom of an old two-liter soda bottle. The bottle should contain at least two or three inches of the mixture. Attracted by the vinegar, the flies will go into the bottle, get caught in the liquid, and drown.

   b. Place one cup of the mixture in a resealable sandwich-size plastic storage bag. Cut a hole about the size of a quarter two inches down from the zipper. Attach the sealed bag with clothespins to an old coat hanger and hang the bag over or near the bin. Again, the flies go in, but they don't come out. Discard when full.

   c. Construct the bag as directed in "method b." Mix one-half cup vinegar with an equal part of water and then add a drop of soap. Next, add a piece of ground meat (the size of a quarter) to the liquid. The meat draws adult houseflies and wasps quite well.

d. Just set out a plain bowl with the vinegar/soap mixture and you will catch flies. Add to the attractiveness of the vinegar by using a yellow bowl. Fruit flies are attracted to yellow.

FLY TRAP #2. This trap is for pesky gnats. Place a glass with two inches of mint mouthwash in it next to the bin. Attracted to the peppermint extract, the gnats will fall into the glass and drown.

FLY TRAP #3. Cut a plastic two-liter soda bottle in half. In the bottom half place a cup of beer or a mixture of brewer's yeast and water. Take the top half, throw away the cap, and place the top half on the bottom half so the spout is pointing down toward the beer. The spout should not touch the beer, but clear it by at least an inch or two. The two halves should fit snugly. (Place a piece of tape around the bottle if they don't.) The flies will smell a yummy dinner and fly in, never to return.

6. The addition of beneficial nematodes to a worm bin will produce good results. These nematodes are selected to attack mostly insect larvae and won't harm your worms or the beneficial organisms living in your bin. Nematodes are usually only available in the spring and summer months, which is the exact same time when flies are usually at their worst. Ask nurseries or home centers for beneficial nematodes and you will get a small box with nematodes mixed in a clay base. Simply follow the package directions for applying. They are not cheap, but a few nematodes is all you need in the worm bin. Use the rest on lawns, gardens, or large outdoor soil worm beds. For more information concerning nematodes, check chapter 4.

7. Have the maggots identified. This will tell you exactly which fly is becoming a problem. Maggots all look pretty much alike, so you will need to take a few maggots to the county entomologist, agricultural-extension office, or local university to help with identification of the fly species.

8. Get a toad. If you live in an area where there are plenty of toads, have one spend the night in your worm bin.

Place wet newspapers down over your bedding. By morning most of the flies hovering in the bin should be gone. But toads will eat worms too, so take the toad back outside the next day.

9. Get out the hand-held vacuum cleaner. It works great to suck up the little flies that are hovering around. (It will also take care of those pesky little fruit flies hovering around the fruit bowl.)

10. When all else fails and the maggots are taking over, changing the bedding is probably the only option left.

A little about flies

Flies are members of the Diptera class of insects. Diptera means "two wings"; all other flying insects possess four wings. All flies have spongy mouth parts and cannot eat anything solid. In order for the fly to eat, it first must dissolve the solid food into liquid form. To do this, the fly deposits saliva, or sometimes its whole stomach contents, onto the solid food it wants to eat. The saliva partially digests the solid food, dissolving it, and then the fly sucks up the dissolved food. This is how a house fly passes germs.

Some flies are looking for the right medium, usually decaying fruit or meat, in which to lay their eggs. This means that the baby flies (maggots) will have a plentiful food source when they hatch.

GRUBS

Grubs have been described as looking like "cocktail shrimp" living in compost piles. Sometimes they can also be found in worm bins. Although these beetle larvae are not pretty, they really aren't all that bad for your bin.

Grubbing around

Grubs are the larval or infant form of various beetles. They are C-shaped and can range in size from a half inch to two inches. In this stage, the grubs are feeding on fresh plant material. They help to break down larger pieces of material into smaller ones that can be eaten by worms and other organisms. In an outdoor worm bed or compost pile, many worm growers do not mind the presence of a few grubs.

Grubs can be accidentally introduced into the worm bin if you feed your worms fresh grass clippings. June beetles and green fig beetles are two beetles associated with grass or sod. Pre-heating the grass, as described in the ant section of this chapter, can prevent grubs from entering the bin. It is also easy to place a melon rind on top of the bin and wait for the grubs to go to it, then toss out the rind and the grubs. For just a few grubs, it works well to hand-pick the grubs out when you see them. Give the unwanted grubs to neighbors with birds or chickens; they also make a great treat for ducks at the neighborhood pond. Your own backyard birds, like jays, small hawks, and road-runners, would also enjoy a nice grub meal.

MILLIPEDES

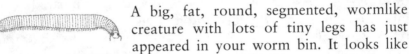

A big, fat, round, segmented, wormlike creature with lots of tiny legs has just appeared in your worm bin. It looks like no other worm you have ever seen. Could this creature be harming your worms? Never fear—it is harmless.

A thousand legs

Even though the millipede's name means "thousand legs," it actually has only about 400. Millipedes, which are arthropods, are cousins of the centipede and belong to the class Diplopoda. Each of their segments has two pairs of legs. Most millipedes range in length from a half inch to two inches, but the smallest is only two millimeters, while the largest tropical millipede can be up to twenty-eight centimeters long. Of the approximately 8,000 described species, most millipedes are harmless, slow-moving animals that are found in moist soils and plant debris. While only a few species are true garden pests, the majority are beneficial, eating dead and decaying plant material.

Millipedes in your worm bin are looking to eat the organic plant wastes, just like your worms. Like grubs, millipedes break down large pieces of plant material for other decomposers. They can usually be found in your worm bin if you use manure or composted plant material for bedding. They aren't true diggers, but will push their way through soft loose soils or bedding. Even though millipedes are harmless, some people do not like the looks of them. They are easily removed from your worm bin by hand-picking.

MITES

 One day you'll open your worm bin and notice thousands of tiny dots moving on the surface of the bedding. What are these dots and how did there get to be so many of them? You're sure they weren't there yesterday.

The small but mighty mite

Mites are very small (about one-fiftieth of an inch) and usually go unnoticed until their populations are in the millions. Close cousins to spiders, mites are arthropods that belong to the order Acarina, the same order to which ticks belong. All mites have eight legs and thus are not insects. (Insects have six legs.) They can be found in almost every habitat in the world, including polar regions, deserts, and hot springs.

There are many different kinds of mites and they eat many different things. Some mites, like white and brown mites, eat only decaying plant material or injured worms, while others eat manure and other mites. Still others are parasites; people in the Midwest and South are familiar with certain mites called chiggers. The red mite you might find in your worm bin is parasitic on earthworms and is definitely a bad guy.

Most people don't realize that mites reside in all worm beds. In well-maintained bins, mite numbers can be kept in control with conditions that are good for the worms but not so good for the mites (see below). When wet and acidic conditions build up in your worm bin, so will the numbers of mites. When mite populations are high, worms will not come to the surface to feed. This deprives the worms of needed food and causes them stress.

Preventing a mite takeover

- Keep conditions good in the worm bin—a pH of 7 is ideal.
- Don't overfeed. Overfeeding is one cause of acidic conditions. Worms eat less in colder months; modify feeding amounts if necessary.
- Mites love wet conditions, so keep the bin from becoming too wet. Keep drainage holes open, feed a drier food, and turn the bed if necessary.

Getting rid of mites

There are several things you can do to reduce the number of mites in your bin. You won't remove them all, but the number of mites will be under control. Of course, if conditions are right for mites, you'll have another outbreak. Mites can be rather difficult to control. Try one or more of the following methods for removing mites.

1. Expose the bin to sunlight. If your bin is small and portable or outside, open the lid to the sun for several hours.
2. Reduce the amount of water and food. This reduction of moisture and food will make the bin environment unfavorable to mites.
3. Place moist newspaper or heavy cloth on top of the worm bedding. Mites will accumulate on the paper or cloth. Remove the paper or cloth the next day and you'll remove a lot of the mites. Repeat as necessary to control mite populations.
4. Place melon rinds or potato slices on top of the worm bedding to attract the mites. Mites will flock to the food, which can then be removed and discarded. Remember, worms like melon rinds, too.
5. Water your bin heavily, but do not flood it. The water will cause the mites to move to the surface of your bin. Now you have several options: Use a hand-held torch to burn off the mites or a strong vacuum to suck them up. If you use the vacuum, place a couple of tablespoons of

boric acid or borax in the bag or holding container. This will kill the mites in the bag. Another option is to sprinkle sulfur on the surface of the bedding to kill the mites. Apply a rate of one-sixteenth of an ounce per square foot of bed surface. This will work well at killing the mites, but sulfur may increase acidity—the exact condition you are trying to avoid.

6. Change the bedding. In extreme cases this may be your only answer.

NEMATODES

Nematodes are unsegmented worms. They are roundworms that belong to one of the largest phyla, Nematoda, which includes more than 10,000 species. This phylum contains some of the most widespread and numerous of all multicellular animals. Nematodes can be found in every environment, in every region of the world. They also can occur in tremendously high numbers. For example, one decomposing apple found on the ground of an orchard contained 90,000 nematodes.

Nematodes have slender, elongated bodies that taper at both ends. They do not have segments like earthworms but instead are smooth-bodied. Nematodes can range in length from less that one millimeter for terrestrial and freshwater forms to over five centimeters for marine species.

Nematodes can be free-living or parasitic. Many parasitic species cause severe crop damage. Some of the nematodes that cause damage are: the stem eelworm, the stinging nematode, and the burrowing nematode. They infect underground bulbs of onions and other root crops. They can survive freezing temperatures, extreme heat, and drying out, making them difficult to control. Crop rotation and organic fertilizers can lead to a reduction in nematodes. Some studies have shown that earthworms can reduce the number of harmful nematodes in the soil.

Nematodes also infest domesticated animals and even man. Diseases and conditions caused by nematodes can range from mild to very severe. Virtually all groups of animals and plants are vulnerable to parasitic nematode attack. For example, the condition elephantiasis is caused by nematodes in the lymph system. It causes swelling in the arms and legs by blocking the lymph vessels. The nematode phylum is one of the most important of the parasitic animal groups.

Nematode species can also be beneficial, especially some of the carnivorous species, which feed on small organisms and even other nematodes. Some nematodes can be applied to a yard to help control the population of insect grubs in the lawn. Others parasitize flea larvae and reduce flea populations. Over two hundred pest species of insects can be controlled by applying nematodes.

The two common species of nematodes commercially sold for garden application, Sc (*Steinernema carpocapsae*) and Hb (*Hererorhabditis bacteriophora*), are available in the spring. Some people use these nematodes to control grubs in their compost bins with good success.

Several nematode species feed primarily on dead organic material and associated bacteria and fungi. These nematodes are an important part of a decomposition cycle and will be found in large numbers in any composting system.

SLUGS

They climb into your worm bins and there they are the next time you open your bin. What are they doing?

Looking for a home

Most of the time, slugs climb into worm bins to look for shade, shelter, or food. Slugs, like worms, need protection from the heat of the day and come out at night to feed on your favorite plants. Most slugs prefer a living plant meal but would not turn down some of the tasty vegetable kitchen wastes you place in the bin.

If slugs are a problem or a nuisance, try one of the following:

1. Hand-pick them out of your bin and drop them into a bucket of soapy water.
2. You can also place shallow pans of beer out under their favorite plants—this will catch quite a few. Check these traps often.
3. Sprinkle diatomaceous earth around the legs of your worm bin. Slugs and snails that cross it will die.
4. Make sure food is well buried.

SOW BUGS

 Most people can recognize sow bugs right away. These creatures are not bugs at all; they're land-dwelling crustaceans that think your worm bin is the perfect home.

A vegetarian isopod

Sow bugs are members of the Isopoda order of crustaceans. They are also called pill bugs, roly-polys, and wood lice.

Most people haven't thought too much about sow bugs, unless they have tried to grow strawberries. Sow bugs can cause a lot of damage to the developing fruit. But, for the most part, sow bugs are beneficial, eating dead and decaying plant material.

Sow bugs also need a moist environment in which to live. Like their lobster cousins, sow bugs have gills, and, without an environment that keeps the gills moist, sow bugs will die. The conditions that are right for your worms are also right for sow bugs.

Indoor worm growers who feed kitchen vegetable wastes probably won't ever find a sow bug in their worm bin, but, if you feed composted manure, you surely will. They generally stay on top or squeeze into cracks and feed on the organic material. Sow bugs won't harm your worms, but if you want them out of your bin, try:

- Sprinkling a little diatomaceous earth on the top of your bedding. (Be sure to use food grade earth.) The diatoms cut the exoskeleton of the sow bug, killing it.
- Hand-pick them out. Don't worry—they won't bite.

SPRINGTAILS

 You see some tiny creatures (one to three millimeters long) that look like white dots. Maybe mites? Maybe not. If you try to touch these creatures and they jump away, then they are probably springtails.

Primitive insects

Springtails get their name from their jumping ability: most of them have a forked structure that folds under them, and when they are alarmed they straighten this structure out, which propels their jump. They belong to the insect order Collumbola, with over 1,200 described species worldwide.

A large springtail that is five or six millimeters in length, can jump up to three to four inches. That would be like a person jumping seventy-five to one hundred feet.

Springtails are usually found living in the soil, under leaf litter, in decaying wood, and in fungi. However, they are not limited to soil situations. Some species can be found living on the surface of ponds, and a few even live in termite mounds. Most soil-inhabiting springtails feed on dead and decaying plant matter, fungi, and bacteria. Others feed on arthropod feces (like millipede and centipede dung), pollen, and algae.

Springtails are very beneficial in the production of soil humus and are considered an extremely important soil organism. Conditions in a well-maintained worm bin are perfect for springtails. They like the nice moist environment and really don't do any harm to your bin. However, some worm growers think they look unsightly in an indoor bin. If you are bothered by the springtails, use your hand-held vacuum to suck up a bunch of them.

7

Using Worms, Castings, and Vermicompost in the Garden

Vermicompost, which is what you get when food scraps and bedding are processed by worms, is one of the main reasons people start keeping worms. To some gardeners, using castings or vermicompost to fertilize the vegetable garden completes the cycle. The gardener has planted the seed and watched it grow. Then the vegetable is harvested and eaten. The trimmings of the vegetable are fed to the worms and the vermicompost that is produced is then replaced in the earth, to wait for the next seed. Gardeners also don't mind at all that vermicompost will outperform commercial fertilizers! Castings are biologically active, and using castings will replenish many of the soil microorganisms needed for healthy soil and plants. Commercial fertilizers can't do that.

It can take a gardener up to 240 days to make a fine grade of compost. But worms can make vermicompost in just thirty days.

There are several ways to use worms and worm products in the garden. Worms can be placed or encouraged directly in the soil, where they can build the soil. Worms can also be kept in bins and their castings used to improve the soil. Water that drains from worm bins can also be used in the garden and on houseplants as fertilizer.

EARTHWORMS IN THE GARDEN SOIL

Earthworms are excellent soil builders. They are generally found in the top twelve to eighteen inches of the soil, where food is most abundant. Worms can enrich a poor soil in many ways: mixing it with their castings, aerating the soil by tunneling, encouraging water retention, and providing microbes and nutrients plants need. Earthworm activity depends directly on soil moisture and temperature. Most worms are active when the temperatures are mild. They move deeper into the soil and become less active in the summer and winter months. By enhancing conditions in your garden, you are putting out the "Welcome Sign" to earthworms.

> Worm mucus has pheromones in it. These pheromones call more worms to join them.

How to invite earthworms to your garden:

• **Mulch it.** Mulching adds organic material to the soil, just like manures and crop residues do. This creates plenty of worm food and helps regulate soil temperature. These are ideal conditions for worms, so the worms will migrate to your garden. This method takes time, but the benefits are worth it.

• **Leave the catcher off your mower.** Like mulch, grass clippings provide food for worms. In a year you could have worm densities of up to 5,000 worms per cubic meter. Clippings can raise the pH of your soil, but with the addition of a little garden lime or calcium carbonate (if needed), the worms and you can have a beautiful lawn.

• **Add worms to a compost pile.** Worms will generally migrate to a compost pile after the initial heating process of the compost. By adding extra worms after heating, you will improve the quality of the compost and speed up the composting process. When it is time to use the compost in the garden, some of the worms will be transported to a new area along with the compost. Soon, worms will be thriving throughout your garden, performing their tasks and building your soil.

• **Add worms directly to your garden.** This usually involves larger numbers of worms than the compost method. Here you add worms directly from your bin to the garden soil. On top of the worms, add at least two inches of good compost. Keep the soil moist so the worms do not dry out. You will need to continue to feed the worms in the soil. Apply another two inches of compost every six months to ensure a good food supply for the worms. Some compost worms, such as *Eisenia fetida*, are not happy unless the soil is rich in organic material. Extra composted manure and organic material like leaves and mulches will be necessary to make them happy. Keep in mind that not all of the worms you place in the garden will survive, but the ones that do will add castings to the soil, and the ones that die will become nitrogen-rich fertilizer.

• **Fertilize it.** It has been shown that organic fertilizers added to the soil result in increased plant production and a simultaneous increase in the earthworm population. The fertilizer increases the soil's organic material by increasing plant production and providing food for the worms.

• **Don't over-till the soil.** Research shows that soils that have been tilled frequently don't have as many worms as soils that have never been tilled. Nightcrawlers, which have permanent tunnels, especially don't like soils that are tilled. Instead of tilling, let the worms do the work. Worms have the capacity to mix the soil and bring up nutrients from lower soil levels. In soils where there are known worm populations, aerating the soil by lifting it with a garden fork may be a better idea.

• **Add garden lime.** Earthworms prefer a slightly alkaline soil, so the addition of a little garden lime to acidic soils will encourage them.

• **Keep soils moist.** Earthworms need moist soils to breathe, so they will leave soil that is too dry. Keeping the soil moist and providing enough food encourages earthworms to stay.

VERMICOMPOST, CASTINGS, AND WORM TEA IN THE GARDEN

The humus that is produced in vermicomposting not only improves the soil, but stimulates plant growth and helps control diseases that attack plants. This humus can be added to the garden in the form of castings or vermicompost. (Vermicompost is a mixture of castings and bacterially composted bedding and food.) One study found that replacing basic plant compost with vermicompost resulted in a 36 percent greater crop yield the first year!

THE NUTRIENT CONTENT OF EARTHWORM CASTINGS IN THE STUDY

1.5–2.2 percent nitrogen

1.8–2.2 percent phosphorous

1.0–1.5 percent potassium

65–70 percent organic matter

The nutrient content of castings varies depending on the type of food eaten by the worms. However, castings have a time-release quality that makes them superior over ordinary compost. When castings were used in the study, the level of nutrients in the soil available to plants was stable over five years! So, let's get some of this stable nutrient material in our gardens.

Adding vermicompost in the garden

• *Work it directly into the soil.* This is easy: just sprinkle it on, rake it in, and forget it. Vermicompost will not burn your plants, so it is worry free.

• *Add some to transplants.* When setting out those small garden seedlings or transplanting mature plants, place some vermicompost or castings directly in the hole the plant is going in. This will give the roots a source of nutrients for growth.

• *Use it as a top dressing fertilizer after the plants have been established.* During the growing season, your plants may need some extra fertilizer. Place one-fourth to one-half inch of vermicompost around the root area or drip line of plants, shrubs, and trees, to give them the benefits of vermicompost.

Using castings in the garden

Castings differ from vermicompost in that they don't have the bacterially composted bedding and food present. In vermicompost, the worms work and rework the material until it becomes dark and earthy-smelling and looks like topsoil.

Pure worm castings still have many nutrients for plants, but the organic material has been broken down further. This can cause the production of salts to occur, which can harm some plants. For this reason, castings shouldn't be used as a sole potting medium. When mixed into soil or potting mixes, castings provide the necessary nutrients while diluting the salt concentrations. Pure castings can also have a higher pH, so some acid-loving plants may not tolerate a higher concentration of castings well.

Many gardeners prefer to work with pure castings, so they know exactly how much casting material to add to plants. Adding as little as 5 percent castings to potting soil can increase the growth and vigor of plants. Castings are nitrogen rich, and the carbolic acid in them also oxidizes with calcium to provide many plant nutrients. Castings may also work as hormones to promote better plant growth.

Castings have been shown to transmit to plants and animals certain resistances to diseases as well. This is called an "acquired resistance system." Two examples of acquired resistance systems are: Chicks were shown to be more resistant to salmonella infections when they were fed castings as part of their diet. Avocado trees have also been shown to resist avocado root rot when castings were used around the tree.

Several potting soil and seed-raising recipes suggest a casting ratio of 25 percent castings to 75 percent potting soil or aged compost. One commercial potting mix includes two parts aged compost, one part castings, and one-half part vermiculite. This is where experimenting a bit can help. Try adding different amounts of castings, say 5 percent, 15 percent, and 25 percent, to potting soils and grow some plants. See which ones do better. If your plants do just as well in the 5 percent casting soil as they do in the 25 percent casting soil, then you can use fewer castings and achieve the same results. This saves more castings to use on other plants. Castings can be used in much the same way as vermicompost—just remember that they are more concentrated.

• *Use castings as a top dressing fertilizer.* In this situation, the castings can become overly dry and form a crust on the surface of the soil. This could slow water penetration. So, castings should be watered well and kept a bit moist to prevent them from drying out. Castings can be used in place of many fertilizers for houseplants. Simply sprinkle a small layer of castings on top of the soil of the potted plant and water well. Repeat the addition of castings every two months or so during the growing season.

• *Sprinkle over lawns and garden soil.* Use castings as a soil amendment for added plant nutrients. Sprinkle evenly on the soil surface, about 1/4-inch thick.

• *Add to transplants.* Place some castings in seed rows or transplant holes. The combination of deep watering and the soil added to the transplant hole will dilute any salt concentrations. Mix equal parts of potting soil and castings and use this when transplanting houseplants into bigger pots.

• *Use as a mulch.* Spread castings about two inches thick within a tree's drip line, staying away from the trunk.

Worm tea

Worm tea is water that has diluted castings in it. This tea can come from the worm bin as drainage or from harvesting castings with water.

To use on potted houseplants, dilute the tea with an equal part water before using. Worm tea can really bring old potting soils to life. For use on plants planted directly in the ground, it is fine to use full strength. Worm tea is especially good for weak or diseased plants. Roots of stressed plants have shown increased vigor when exposed to worm tea. Drench stressed plants with worm tea or use it in a standard feeding program.

UNWANTED HOUSEGUESTS IN YOUR POTTING SOIL

When you use vermicompost and castings in your potting mixes, sometimes the associated organisms also get transported to your houseplants. Potting soil mixed with vermicompost is alive with soil-building microorganisms that are beneficial to the soil of any living plant. However, some people draw the line when it comes to knowing these organisms are in their homes. Springtails and mites are just two examples of unwanted guests. Since most people associate mites with harming their plants, they want them gone.

If you really need to know they are gone, try heating your potting soil in the oven. Put the soil in an old roasting pan and heat it at 200°F for thirty minutes. This should kill most of the offending creatures.

Remember, though: Most creatures in the worm bin feed on dead organic matter, and if there is no dead organic matter to feed on in your houseplants, the organisms will die. So don't worry if you see a few going into the pot. For more information concerning pests, associated with worm bins, check chapter 8.

WINDROWS

Windrows are worm beds in the ground. Many worm growers use this method because they feel the beds produce a better quality of worm. The beds usually have good drainage and aeration, and food is always available for the worms. If conditions aren't just right, the worms have someplace to escape to until the conditions are favorable again.

ONE EASY WINDROW SYSTEM WORKS LIKE THIS:

Spread about six inches of aged composted manure in a row. The row can be as long as you like, but the width should be no more than two to three feet wide. This way you can tend across the row as the windrow gets bigger and bigger. Some large commercial windrows are wider across, but you should do what is comfortable for your back. Now, seed the row with worms, using one pound of worms per linear foot as a guide. Use a sprinkler system to keep the windrow moist. Every one to two weeks (depending how much compost is left), add another four to six inches of aged composted manure on top of the row. When the windrow gets to be three to four feet high and three feet wide, it is time to harvest. Take off the top layers (most of the worms will be found in the top anyway) and leave one foot of vermicompost as the base. Re-seed and add more composted manure. Repeat.

We have heard of worm growers using layers of straw and fresh manure as food as well. The idea is that the straw separates the fresh manure from the worms until the manure composts and decomposes the straw, too. You may want to experiment with this idea.

Shrews and moles

In areas that have shrews or moles, many worm growers complain that the little pests will move in and eat the worms in their windrows. Moles can eat three times their weight in just one day, and they love worms. Shrews eat every hour to keep up their metabolism. So, what are you going to do?

First, determine which tunnels are active and which ones are just for foraging. Stomp on and press down all the tunnels around the worm beds and in your yard. But beware—moles will repair the main tunnels. Moles can dig about two hundred feet per day if necessary.

Now that you have figured out which tunnels the mole is using, you can try several methods:

- Wet the tunnels down thoroughly. Moles do not like very wet soils.

- Place dog or cat feces in the tunnels. The mole will think a predator is nearby and leave. The feces will not remain effective for long, so alternate cat and dog feces if you can.

- Like worms, moles are sensitive to vibrations, so place whirligigs at the edges of your property to deter the moles from coming back. Place the whirligigs well away from your beds.

- Bury fine wire mesh or hardware cloth in the ground before establishing your worm bed. You will need to go down at least eighteen inches to keep the critters out. Carefully secure wire sections that meet to prevent the moles from finding an entrance. A very determined mole may still find a way in, but this will definitely slow him down.

• Moles dislike castor-oil plants, which are very poisonous. They should only be planted with extreme caution, away from children, pets, and worm beds.

A little life history about moles: Moles are not rodents; they are not related to gophers. They are members of the mammal order Insectovora, and eat only insects, grubs, and worms, not plants. They range in size from two and a half to six inches in length, are usually brown to black in color, and have very broad front feet and short tails. Their eyes are very small (about the size of a pinhead) and they have no external ears.

Moles are very territorial and are active both day and night. They are in constant search of food and prefer living in soft soils, like your worm beds. They do not like rocky soils, and you can sometimes deter them by burying a layer of gravel around your bed. A little good news about moles: Because of their need for large quantities of food, it is estimated that only two to three moles can survive per acre.

MAKE A MOLE REPELLENT.

2 tablespoons castor oil
1 tablespoon liquid dishwashing detergent
1 gallon water

Mix all ingredients together and wet down the mole's tunnels with the solution. Keep solution away from worm beds.

8

EARTHWORMS IN AGRICULTURE

Earthworms play an integral part in agriculture. Where once earthworms built up the soil naturally, the use of modern farming practices has greatly reduced their numbers. Some will say that through chemical means, like fertilizers and pesticides, we have increased crop production. This is true, but each year our soil is becoming more and more barren. It takes more and more chemicals to grow the same crops. The soil microorganisms that help produce humus and give soils their growing capacity are dwindling. Things need to change, and an example of forced change is the small country of Cuba.

> If we used all of nature's resources except earthworms to produce topsoil, it would take us about 100 to 150 years to produce one single inch! Bring back the earthworms, and it only takes one year to produce that inch.

In 1986, the Cuban government started a vermicomposting program to design safe and effective soil management tactics. Cuba was caught in a vise of economic sanctions, political pressures, and lowered crop production. Cuba was faced with no choice but to find alternatives to its past dependency on imported fossil fuels, fertilizers, pesticides, and animal feed.

Cuban scientists developed a full technological package for the production of humus from earthworms, a process known as vermicomposting or vermiculture. They found the best application rate was four tons per hectare of earthworm humus for most crops. Since the implementation of this program, imported agricultural products have been cut by as much as 80 percent.

Cuba's vermicomposting program started with two small boxes of redworms, *Eisenia fetida* and *Lumbricus rubellus*. Today there are 172 vermicompost centers throughout the country. In 1992, these centers produced 93,000 tons of worm vermicompost. Several different institutions and companies are involved in vermiculture operations, but most of the research is conducted by the Institute of Soils and Fertilizers and by the National Institute of Agricultural Sciences.

Cuba's vermicompost program is simple and effective. First, manure is composted for approximately thirty days aerobically, and then transferred to open vermicompost beds. At some sites, like the Pinar del Rio Vermiculture Center, these vermicompost beds serve two functions. The beds are located between the rows and in the shade of large mango trees. The trees are fertilized from nutrients leached from the beds, and worms and vermicompost can be harvested as well.

The beds are approximately one and a half meters wide and are of various lengths. The composted manure, which is mainly from cows, is mixed with soil and "seeded" with earthworms. Most vermicomposting operations in Cuba use cow manure as the primary source of organic material. Some other sources of organic material for the beds are: pig and sheep manure, filter press cake from sugarcane, coffee pulp, plantains, and municipal garbage. After set up, the vermicompost beds are monitored to maintain proper moisture and temperature requirements.

The worms are fed the composted material, which is placed on top of the beds. The worms then deposit their castings in the lower levels. Compost is continually added until the beds reach a height of approximately one meter. This usually takes about ninety days and then the worms and castings are harvested. The worms are concentrated in the top ten centimeters of the pile and are scraped off or separated from the vermicompost in a screening process. The vermicompost is either dried and bagged or used on-site as a soil amendment and fertilizer.

In Cuba, even animal feed depends on worm production. Worms that are not placed back in the vermicomposting beds are dried and chopped and added to animal feed as a protein source. Earthworms are made up of 75 percent protein, with little fat, and they require very little processing. This makes them an excellent protein additive for animal feed.

In other countries, including the United States, fish by-products are often added to animal feed to provide additional protein. During years when the fish catch was poor, the commodities market sees higher prices for animal food, due to the higher cost of fish. This in turn raises prices for the consumer. Maybe Cuba is leading the way for a new alternative protein source: earthworms.

Cuba isn't the only country looking to worms to improve soil conditions. The Ecology Institute of Mexico, Peru's INIAA/NCSU Yurimaguas Experimental Station, the University of Rwanda, and Spain's Universidad Complutense have all completed the first stage of earthworm projects. India, France, and Australia have been studying the benefits of worms for quite some time and have vermicomposting facilities. Most of the countries in the world are taking a serious look at the impact of earthworms. Vermiculture is definitely in its infancy throughout the world, but it is growing rapidly.

CROP MANAGEMENT IMPACT ON EARTHWORMS

When we manage soils for crop production, we in turn are also managing the habitat of earthworms and soil organisms. By managing the crops with modern agricultural means, we are directly affecting the habitats where earthworms live. Farming practices impact a worm's food supply by removing much of the soil's organic material or by changing its quality so it's not suitable for the worms. A worm's environmental conditions can be altered by removing the mulch protection of the soil that allows fluctuations in moisture and temperature levels. The soil's chemical condition is greatly affected by the use of chemical fertilizers and pesticides, making the soil inhospitable for worms. By considering how these factors affect worms in different crop managing systems, we can start to predict the effects on worm populations.

Studies have been conducted on several different crop management systems over a ten-year time span. The studies indicated that management practices that included the addition of organic material to the soil, along with little tillage, produced the highest numbers of earthworms. On the reverse side, management practices that included no addition of organic material and severe tillage showed the lowest numbers of earthworms. When farmers plow their fields, more and more earthworms that naturally till the soil die, so more and more plowing is necessary. This is a vicious cycle in our large mono-crop system. Every inch of soil a tractor travels on is compacted a bit more, so the farmer plows deeper and deeper.

The impact of agricultural chemicals on earthworms varies with the chemical used. Inorganic fertilizers added to fields promote greater plant production than occurs in fields that are not fertilized. This in turn promotes higher earthworm populations. It has also been shown that most herbicides are only slightly toxic to worms—a surprising outcome. When other pesticides were tested, they usually fell into two categories: Harmless to moderately toxic, and highly toxic. Some highly toxic pesticides had milder effects when only spot treatments were done. In general, organophosphate and pyrethroid pesticides fell into the category of harmless to moderately toxic; while carbamate pesticides, fungicides, and nematicides were highly toxic to earthworms.

ENCOURAGING EARTHWORMS IN CROP MANAGEMENT

Now that we know how earthworms are affected by crop management, the problem is finding a way to reverse the decline of earthworm populations. Following are some possible solutions:

1. *Reduce tillage.* This leaves more surface mulch for food, and provides cover to protect the worms from environmental conditions.

2. *Grow winter crops.* These crops will add to the mulch protection, in addition to providing additional food for the worms.

3. *Add organic material.* Fertilizing with composted manures will greatly increase the organic portion of the soil and encourage worms to stay.

4. *Rotate crops*. Rotating crops with hay or fallow fields will provide more food for the worms.

5. *Maintain a pH balance*. Maintaining a soil's pH between six and seven provides optimum conditions for the worms. It should be noted that the proper pH isn't always a guarantee that earthworms will move in. For example, if the soil's texture is too coarse or there are drainage problems, earthworms might not want to live there.

6. *Use pesticides sparingly*. Using spot treatments or a pesticide with very low toxicity will help keep worm populations up.

SHOULD FARMERS "SEED" THEIR FIELDS?

This is an interesting question being asked by many farmers. If crop management practices haven't changed for quite some time, then a given field already holds as many earthworms as the soil can support, and adding additional earthworms isn't going to help. Most of the worms will die and their bodies will become a bit of fertilizer, but that's about all the benefit you are going to see.

However, if a newly planned crop management system will include many features that encourage earthworms, then some "seeding" could be beneficial. As soon as conditions change in favor of earthworms, they will move in naturally, but this could take one to two years. Many farmers have claimed success at establishing earthworms after conditions become more favorable.

One inexpensive way to establish earthworms is to collect native earthworms yourself from pastures and roadsides on rainy spring nights or early mornings. This way you will know that the worms are adapted to the native soils and climate. Once back at the field, place four to six earthworms together under a small pile of mulch. Do this every thirty or forty feet, on a cloudy, wet, or cool day. Record where you "seeded" the worms and check periodically for worm activity. Hopefully, the worms will survive and spread. Unfortunately, this can be a very labor intensive method and may therefore be impractical for people with large farms.

9

COMMERCIAL WORM GROWING: CAN YOU DO IT?

Worm farming is said to be the fastest growing agricultural industry in the country. Worms are making millionaires out of people overnight. Well, that may be an exaggeration, but there are some pretty big claims being made about the easy money to be made from worms.

In this chapter, we will take a look at these claims and look at the many possible ways worms and worm products can be turned into a money-making business. We will look at various aspects of different businesses, so you can decide if starting a worm business is right for you.

SELLING WORMS THEMSELVES

Producing worms for bait, vermiculture, agriculture, and home composting are several ways that a business can revolve around worms. In these business ventures, the end product you are selling is the worms themselves.

Most people are quite familiar with growing worms for bait. However, in Canada there is an industry for collecting worms in the wild. Worm pickers go out at night after a rain and pick up nightcrawlers. They strap cans to their legs, wear mining lights on their heads, and head out to their favorite hunting grounds, such as open fields and golf courses. It is estimated that a good worm picker can make up to $150 to $200 per night. The competition is fierce, and there have been several reported instances of fights between groups of worm pickers that have all arrived at the same spot. Most of these worms are exported to the United States. Nightcrawlers are very difficult to grow because of their permanent burrows, and attempts to raise them commercially have failed. Nightcrawlers are a large market of about $100 million annually at present. Many bait growers rely on growing redworms, but this is a much smaller market, totaling only a few million dollars. Bait worm growers usually provide worms to local fishing areas.

Supplying worms for vermiculture and home composting systems is a growing field. With the price of worms at between $8 and $30 per pound in the retail market, this sounds pretty good. One problem with this is that there is no longer a need for the customer to buy more worms after the initial shipment, since worms can reproduce quickly. This venture then becomes a search for more and more new customers, with very little repeat business.

Growing worms for agriculture is new and basically just starting out. Cuba grows worms as its own protein source for livestock feed. This could be done here as well. Worm protein is comparable to fish meal protein, which is currently used for livestock feed. Fish meal prices vary with how much fish is caught in a season: Sometimes the fishermen catch a lot, and some years the catch is low, which drives prices higher. With worms you can count on a steady supply. The only preparation needed for shipment is that the worms need to be dried and sold in bulk tonnage. It is estimated that dried worms could find a very large market.

Zoos, pet stores, and hobbyists who own birds and reptiles all need a supply of worms. Becoming a distributor to these businesses and individuals can prove profitable.

SELLING CASTINGS

Scientific research has documented the benefits of castings as a soil amendment and as a benefit to plants. They can be sold in small bags or in bulk: pure castings can sell for anywhere from $2.50 to $5.00 per pound retail. Many organic gardeners and farmers appreciate a vermiculture way of recycling and fertilizing, so try contacting any organic growers associations in your area and talk to them about the idea.

RECYCLING GREEN WASTES WITH WORMS

The landfills are almost full, and many counties and cities are looking for ways to reduce the amount of waste going to landfills. Vancouver, Canada, and San Jose, California, are just two municipalities that are promoting waste reduction by encouraging vermicomposting at home. Several worm growers who have taken this idea and developed cooperative agreements with trash companies are now recycling green wastes into castings and vermicompost. The worm grower sometimes receives a collecting or processing fee for the service. Other worm growers are happy to receive the unlimited raw material and make their money in the sale of worms and castings.

MAINTAINING VERMICOMPOSTING SYSTEMS
FOR SOMEONE ELSE

With more people starting up vermicomposting systems, a need arises for a vermicomposting system troubleshooter or a person to do the harvesting for people who want a bin but not the work.

The "vermicomposting expert" could be called upon to set up, maintain, harvest, or correct problems in a bin. With more and more schools and institutions opting to reduce their disposal costs, starting a vermicompost bin system on site becomes necessary and practical. Also look to other businesses that produce quantities of waste, like dairies and horse ranches. In all of these cases, both you and the business could greatly benefit from vermiculture.

MARKET AND DISTRIBUTE VERMICOMPOSTING PRODUCTS

As the idea of vermicomposting grows, so does the need for worm bins, sorters, gauges, and equipment of all kinds to take care of the systems. New automated systems and easy-to-use testing instruments need to be invented, marketed, and sold.

Start a newsletter or get on the Internet and "shop" your products. Provide needed products to solve the vermiculturist's current problems.

START A VERMICOMPOST DELIVERY BUSINESS FOR LANDSCAPERS

As the benefits of vermicompost and castings become better known to landscapers and homeowners, the requests for these products will increase. Offer to deliver castings and vermicompost to work sites and home backyards. Provide information as to their uses and application rates. Provide the expertise or labor—or both.

CONDUCT VERMICOMPOST WORKSHOPS

Teach other people how to maintain a successful worm bin. Write educational pamphlets and conduct demonstrations at schools, garden clubs, fairs, and expos. Spread the word about vermicomposting. Make videos with step-by-step instructions or one for children. The ideas are endless. If you don't want to write the material yourself, then publish or distribute material written by someone else.

SELLING WORM TEA

Just like castings or vermicompost, worm tea is an excellent fertilizer. Bottle the tea and sell it with your castings as liquid casting fertilizer. Worm tea can sell for $1 to $3 per gallon.

STARTING YOUR OWN WORM BUSINESS

Once you have decided on what type of business you want and how to go about it, then comes the next step: What do you need to start? The following is a general list for most vermiculturists. Your worm business may or may not need all the following or may need more.

1. Every book, magazine, and article you can find on the subject. Buy this book! Get a feel for what you are about to do.

2. Determine the size of the operation you need. Do you want to start small and grow? Do you have the space for a large operation in your backyard? If your operation becomes large, is your property zoned correctly?

3. Determine the location of your operation. Locate it near a water source, electricity, or maybe in a shed. Where would it function best in your climate and location? Where will it be easy for you to tend?

4. Find suppliers of necessary quantities of bedding and feedstock for your operation. Is there a horse ranch nearby? Paper mill?

5. Obtain necessary tools. General tools include a pitchfork, pH meter, compost thermometer, moisture meter, hand claw tool, and shovel. If you are going to use paper as your bedding source, a paper shredder would be handy.

6. Build or buy commercial bins or prepare a windrow. Depending on the size of operation you wish to have, you will need to build or buy worm bins, or prepare a windrow with the correct surface area for the worms. Remember that redworms need one square foot of surface area for every pound of worms.

7. Obtain bedding and feedstock and do any preparations of that material if necessary. Are you shredding newspaper, or do you need to leach manure?

8. Buy worms. Have everything ready before the worms arrive! What worms did you decide to grow? Get the scientific name.

Now you have started your worm business. You are now on to the next step:

MAINTAINING YOUR WORM BUSINESS

1. Feed worms properly.

2. Monitor conditions in the bin or bed. Check pH, moisture, and aeration. Soon you will be able to tell just by looking if everything is going smoothly, but check anyway.

3. Adjust feeding and moisture during different climatic seasons. The bin or bed that is outdoors will be affected most by the weather. Use less water and feed in the winter and more water and feed in the spring, summer, and fall seasons.

4. Prepare more beds and bins for the first harvest. This may simply involve preparing more bedding to go back into the original bin, or you may decide to divide the worms into two bins.

5. Prepare harvesting equipment. Are you going to hand-screen everything or are you going to purchase a harvester?

6. Package castings or vermicompost.

7. Advertise.

8. Sell castings and extra worms.

9. Repeat process.

Of course, this is an example of a simple vermicomposting procedure. You take it from here. You may decide to conduct experiments with your vermicompost and see how well it works. The only limit here is your own imagination. Good luck!

THINGS YOU CAN DO TO GUARANTEE YOUR SUCCESS

1. Plan, plan, plan. Think your plan out thoroughly. What do you need in order to start? What are your strengths? Weaknesses? What are you good at, and how can you

use worms to start a business you will enjoy? Once you decide in which direction you want to go with your worm business, then look at the problems of money, understanding the market, and putting together a complete business plan. Making a well-thought-out business plan will send you in a clear direction, with few surprises.

2. Keep good records. This is extremely important and often overlooked. Keep a careful record of time, money, and receipts for your new business. Tax laws are complicated, and this thorough record keeping will benefit you in the long run.

3. Make sure you get the proper licenses. A business license is generally required for all businesses, but check your state and county business administrations for additional permits or licenses you may need.

4. Keep learning. You can never know everything, so keep learning every new aspect of vermiculture. What are other people in the field doing, and what is the latest bin design? Learn everything you can.

5. Start your business with sound, proven ideas. Once you are up and running successfully, improve on them. Be creative and find the perfect worm bin. Use your experience with the proven and then be creative with the untried and develop something new.

6. Work hard. No business is easy! Even if it becomes easy to you, do something every day to increase your business or make it better. Never lose sight that a business takes work to keep it growing and thriving.

7. Advertise. Let people know what you are doing. Print flyers and business cards or advertise in the Yellow Pages or on the Internet—just let people know. You could have the customers right next door, but if they don't know you are there and what you have to sell, then you don't have customers.

8. Look for opportunities. Let's say you produce the best castings around and your business is doing well. Would it be better if you sold the castings at a nursery or home

center? Is there an Earth Fair coming to your area that is looking for vendors? These are opportunities to make your business grow, but you must look for them. Vermiculture is closely tied to horticulture and waste disposal, and there are many avenues one can enter into with these industries to improve or expand a business.

9. Be a community expert. Vermiculture is a new field: Who do you turn to if you need help with a bin or worm problem? Well, you can become that person. Give talks to garden clubs, nurseries, and schools. Your business can only benefit from the added exposure.

10. Join professional groups. This gives you the opportunity to find out what is happening now in vermiculture. Professional associations usually have meetings and newsletters to discuss cutting-edge technology and recent studies.

THE INTERNATIONAL WORM GROWERS ASSOCIATION

This new nonprofit organization is aimed at helping worm growers with information, meetings, a newsletter, and future projects to help vermiculturists. Membership information can be obtained from:

Rick Best, Chairman, IWGA
P.O. Box 900184
Palmdale, CA 93543
E-mail address: wormwise@aol.com

SCAMS

Unfortunately, the worm business, like any other, has been the victim of scam artists. We mention it here to remind you to be on the alert for such scams. These business ventures often sound too good to be true, and they usually are.

PYRAMID SCAM:
This scam involves a person who starts a worm farm. Let's call him Con Worm.

Con Worm convinces other people that there is such a high demand for his worms that he can't keep up with the demand. The orders are coming in but he just can't fill them. Con Worm then sets people up with expensive worms and equipment and promises to "buy back" all the worms they can produce so they will make lots of money. The people hand over their money (sometimes thousands of dollars) to Con Worm, and they take their worms and equipment home. At this point the scam can go two ways:

1. The new worm growers can't seem to get it right and don't produce any worms for Con Worm to buy back. Remember, any business is hard work. They blame themselves because it didn't work and go back to Con Worm, who "feels sorry for them" and buys back the equipment for pennies. Con Worm then turns around and sells the equipment for full price to someone else.

2. The new worm growers get it right and start producing worms for Con Worm to buy back. Con Worm buys the worms, all right, but sells them to new people he has convinced to supply worms for him. Now, when he can't convince any other people to be his worm suppliers, he stops buying the worms back and everyone goes out of business.

ANOTHER SCAM:

A worm company offers to sell you worms at a big price and KEEP THEM FOR YOU at their farm. They promise to double your money in a year by selling the baby worms. But, instead of selling the baby worms to valid markets, they sell the baby worms to other people like you, telling them they are going to double their money as well. Eventually, when they can't get any new investors in the scam, they declare bankruptcy and the people who invested lose their money.

So, buyer beware—and do your homework before you invest.

10

THE WHOLE WORM AND NOTHING BUT THE WORM

While doing the research for this book, we found many stories, poems, songs, and art that focused on worms. Some were funny, some wild, and others date back many years to native folklores. Because we thought no one would believe half the stuff in this chapter, we decided a parody of "the whole truth and nothing but the truth" would be perfect.

POETRY

Little did we know that worms had inspired the famous poet Walt Whitman, who wrote a poem called "This Compost" about worms recycling the dead and decaying.

The following is a portion of the poem:

Something startles me where I thought I was safest,
I withdraw from the still woods I loved,
I will not go now on the pastures to walk,
I will not strip the clothes from my body to meet my lover
  the sea,
I will not touch my flesh to the earth as to other flesh
  to renew me.

O how can it be that the ground itself does not sicken?
How can you be alive you growths of spring?
How can you furnish health you blood of herbs, roots,
  orchards, grain?

Are they not continually putting distemper'd corpses
      within you?
Is not every continent work'd over and over with
      sour dead?

Where have you disposed of their carcasses?
Those drunkards and gluttons of so many generations?
Where have you drawn off all the foul liquid and meat?
I do not see any of it upon you to-day, or perhaps
      I am deceived,
I will run a furrow with my plough, I will press my spade
      through the sod and turn it up underneath,
I am sure I shall expose some of the foul meat.

Behold this compost! behold it well!
Perhaps every mite has once form'd part of a sick person—
      yet behold!
The grass of spring covers the prairies,
The bean bursts noiselessly through the mould in
      the garden,
The delicate spear of the onion pierces upward,
The apple-buds cluster together on the apple-branches,
The resurrection of the wheat appears with pale visage
      out of its graves,
The tinge awakes over the willow-tree and the
      mulberry-tree,
The he-birds carol mornings and evenings while the
      she-birds sit on their nests,
The young of poultry break through the hatch'd eggs,
The new-born of animals appear, the calf is dropt
      from the cow, the colt from the mare,
Out of its little hill faithfully rise the potato's
      dark green leaves,
Out of its hill rises the yellow maize-stalk, the lilacs bloom
      in the dooryards,
The summer growth is innocent and disdainful above
      all those strata of sour dead.

What chemistry!

Now worm poems are popping up everywhere: Several museums hold poetry contests with worm and gardening categories. And one of the authors of this book was inspired to pen a worm poem.

WORMS IN MY HAIR
by J. H. Taylor

The worms, worms they are everywhere!
I worked in the bin, now they're in my hair.
Sorting, screening from bin to bin,
Within their casting from rim to rim.
They've turned my trash into heaps worth gold,
With little help from me the old.
My plants will thank them one and all,
For providing nutrients when they call.
You can do it, yes you can,
Turn that trash into black fertile land.

SONGS

Worms have been the inspiration for many songs, and not just the silly ones we all sang as children. Do you remember?

Nobody likes me, everyone hates me, guess I'll eat some wor-r-r-rms,
Chorus:
Long, slim, slimy ones. Short fat, juicy ones,
Itsy bitsy, fuzzy, wuzzy worms.
First you bite the head off.
Then you suck the guts out.
Oh how they wiggle and they squir-r-r-rm.

Down goes the first one, down goes the second one,
Oh, how they wiggle and they squir-r-r-rm.
(Chorus)
Up comes the first one, up comes the second one,
Oh, how they wiggle and they squir-r-r-rm
(Chorus)

> Nobody hates me, everyone likes me,
> Why did I eat those wor-r-r-rms?
> Long, slim, slimy ones. Short, fat, juicy ones,
> Itsy bitsy, fuzzy, wuzzy worms.

For those of you who had gross little brothers, you probably also remember:

> The worms crawl in, the worms crawl out,
> the worms play mumblety-peg on your snout.

Well, many songwriters have also been inspired by the earthworm and several groups have included worm songs on their albums. For example:

Pink Floyd, "Waiting for the Worms," on their album
  The Wall.

Cathedral, "Enter the Worms," on their album *The Ethereal Mirror.*

Comecon, "Worms," on their album *Converging Conspiracies.*

Echobelly, "Worms and Angels," on their album *On.*

Pogues, "Worms," on their album *If I Should Fall From Grace With God.*

Theatre of Ice, "Gone With the Worms," on their album *Murder The Dawn.*

TELEVISION

A lady and her worm bin were part of a story line on *ER.* The Emmy Award–winning show featured worms that had been left out in the cold in a Chicago winter. Does this mean that even worms need to go the emergency room once in a while?

ART

Yes, earthworms have joined the art world as well. They create their own abstract art. Their bodies form the patterns and we humans are the beholders of their creations. No, I didn't make this up! The process involves using worms as the brushes.

To do this yourself, you will need:

Several different watercolors or water-based acrylic paints

About a dozen worms of different shapes and sizes

Canvas or watercolor paper

Dip a worm completely in one color of paint. Drop the worm onto the paper and let it wiggle around a bit. Do this with another worm and another color. You can remove the worms after each color and let that color dry a bit before dropping the next worm, or drop the worm with the next color onto the paper while the last color is still wet and allow them to mix. You decide. When you're finished with the worms, wash them off and return them to the bin.

The final picture will be a definite conversation piece!

"CALLING ALL WORMS" OR WORM GRUNTING

Yes, you can call worms right out of the ground. This practice has been traced back to primitive societies around the world. The art of worm calling has many important uses. Some animals, like kiwis, turtles, and gulls are known to call worms. Maybe by observing and copying these animals, the ancient peoples learned how to effectively call worms also.

Every year the U.S. Forestry Service sells about 700 worm calling permits in the Florida panhandle.

It is pretty easy to call worms. The object is to send vibrations down into the soil, which will cause the worms to come up to the surface. No one really knows why worms will do this. Some say the vibrations mimic the digging sound of moles and shrews and the worms come to the surface to escape them. Others say the vibrations mimic falling rain and the worms come to the surface to mate or keep from drowning (although we now know worms aren't in danger of doing that). Finally, another theory is that the vibrations mimic seismic disturbances and cause the worms to surface. Whichever it is, this would make a fun science experiment!

There are many ways to call worms. One method the ancient people probably used is two sticks. One is smooth and the other

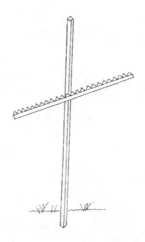

WORM GRUNTING STICK

is notched like a saw. First, you pound the smooth stick at least six inches down into the ground, the further down the better. Now use the notched stick and rub it across the smooth stick in a sawing motion. Do this until the worms surface. (This could take from a couple of minutes to half an hour.) Practice in an area that you know has worms first, like your worm bin. This will help you perfect your technique, so when you are at that fishing hole, you will know your worm calling will work.

Back in our grandparents' time, anything that would send vibrations into the ground was tried, but some things worked better that others. Tapping a metal garden tool like a spade blade, rake, or hoe against the ground worked pretty well. Sometimes they got impatient, though, and the worms were just dug.

Another way to call worms is to pound two metal rods into the ground about two feet apart and connect a car battery by jumper cables to each rod. Worms are said to just jump out of the ground.

You can also try to call worms by driving a galvanized pipe several feet into the ground. Use a pipe that has a pretty large diameter, and, when you want some worms, just place a drain snake down the pipe and turn it on. The snake vibrates in the pipe and the worms pop right up.

In the South, the slang name for worm calling is "worm grunting." But whatever you call it, it is a great skill for fishermen, Boy Scouts, or anyone who needs worms.

11

COOKING WITH EARTHWORMS

Earthworms, by all accounts, are a nutritious addition to human diets. They contain sixty to seventy percent protein and very little fat. Earthworms are entirely edible and, aside from the initial preparation, are very easy to use in recipes. Worms can be boiled and chopped and used in casseroles, or they can provide the main protein source instead of chicken or beef. They can also be baked and used as a flour for baked goods.

Several organizations and universities have sponsored earthworm and insect cooking contests, and everyone involved has enjoyed some wonderful dishes. The earth's food supply has to feed more and more people, so someday a dish called *Soufflé Ver de Terre* may be one of your favorites. (*Ver de terre* is French for earthworm.)

INITIAL PREPARATION

Just like snails, earthworms should be purged before you eat them. After reading this book, you know what the worms could be eating, so you'll understand why purging is a good idea.

If you ordered new worms to cook, it is best to transfer them to moistened corn meal for forty-eight hours as soon as you receive them. This gives the earthworms enough time to purge their systems. After that, the worms can be washed and frozen for later use. One cup of earthworms is equivalent to one-half pound.

To wash worms, rinse them vigorously in cold water. Use a fine colander for small worms. Transfer them to a paper towel and remove any dead worms. (They will be the ones that aren't moving.) Now it's time to cook the worms. The most popular methods of cooking are boiling and baking.

To boil earthworms, start two pots of boiling water. Place the live worms in one pot of boiling water for 15 minutes. Remove the worms and place them in the second pot for another 15 minutes. The object is to remove all the mucus covering the worms. After the second boiling, check the worms and, if any mucus remains, boil them again. Finally, rinse and pat dry. The worms now can be used in recipes or frozen for later use.

To bake earthworms, first freeze them (so they are dead and don't wriggle off the baking sheet), then defrost them. Preheat an oven to 200°F. Place clean, defrosted earthworms in a single layer on several sheets of paper toweling on a baking sheet, and pop them into the oven. After 30 minutes, check the worms. Adjust the time you bake your worms to the recipe's requirements. Baking small worms for 30 minutes should result in very dry worms that can easily be ground into flour.

EARTHWORM RECIPES

Just about any favorite recipe can be protein-enhanced with a bit of earthworm flour. Add a tablespoon or two to cakes, muffins, or breads. Experiment with using chopped worms instead of raisins or nuts. The possibilities are endless!

WORM'N'APPLE CAKE

3 eggs, beaten
1 $^3/_4$ cups sugar
1 cup vegetable oil
1 teaspoon vanilla
2 cups flour
1 teaspoon salt
2 tablespoons earthworm flour
1 teaspoon nutmeg
1 teaspoon cinnamon
1 teaspoon baking soda
3 or 4 chopped apples
1 cup chopped nuts
Confectioners' sugar for dusting

~~~~~~~~~~~~~~~~~~~~~~~~~~~~~~

Preheat oven to 350°F. Beat the eggs, sugar, oil, and vanilla. In a separate bowl, combine the flour, salt, earthworm flour, nutmeg, cinnamon, and baking soda. Combine wet and dry ingredients and beat well. Add the apples and nuts.

Bake in a tube pan for 50 to 55 minutes, or in a 9 x 13 pan for 30 to 35 minutes. Remove the cake from the oven and let it cool in the pan. Remove the cake the from the pan and sprinkle with confectioners' sugar. Enjoy.

12 servings

VERMICELLI WITH EARTHWORM MEATBALLS

Meatballs:

1 1/2 pounds ground beef

1/4 cup earthworm flour

1/2 cup dry bread crumbs

1 tablespoon parsley

1 teaspoon salt

1/8 teaspoon pepper

1 teaspoon Worcestershire sauce

1 egg

1/2 cup milk

1/4 cup oil

1 pound packaged vermicelli

2 (26-ounce) jars of your favorite spaghetti sauce (or your homemade sauce)

Grated Parmesan cheese

Mix all meatball ingredients except the oil. Shape the mixture into 1 to 2-inch meatballs. Place the oil in a large skillet and heat to medium hot. Cook the meatballs until done, about 15 to 20 minutes.

While the meatballs are cooking, heat the spaghetti sauce in a large saucepan. Boil the vermicelli according to package directions. Drain. Place cooked, hot vermicelli in a large bowl or platter. Top with meatballs and sauce. Sprinkle grated Parmesan cheese over all. Serve immediately.

Serves 6 to 8

APRICOT-EARTHWORM BALLS

8 ounces dried apricots, finely minced
2 1/2 cups flaked coconut
3/4 cup sweetened condensed milk
2 tablespoons finely chopped boiled earthworms
1/4 cup finely chopped raisins
Confectioners' sugar

Combine all ingredients except the sugar. Shape into 1-inch balls. If the balls are a bit dry and won't hold together, add a bit more sweetened condensed milk. Roll balls in the confectioners' sugar. Place in a container in the refrigerator until firm, 1 to 2 hours. Enjoy. Keep leftover candy in the refrigerator in a sealed container.

Makes about 4 dozen

OATMEAL EARTHWORM-RAISIN MUFFINS

I egg
I cup milk mixed with I tablespoon vinegar
$1/2$ cup packed brown sugar
$1/3$ cup shortening
I cup quick-cooking oats
I cup flour
I teaspoon baking powder
I teaspoon salt
$1/2$ teaspoon baking soda
$1/2$ cup coarsely chopped boiled earthworms
(or $1/4$ cup each of worms and raisins)

Preheat oven to 400°F. Beat the egg and add it to the milk and vinegar. Stir. Mix the brown sugar and shortening together. Combine the egg/milk mixture with the brown sugar and shortening. Stir in the remaining ingredients, except worms, until the flour is just moistened. Gently stir in the worms. The batter will be lumpy. Grease 12 muffin cups or line them with paper liners, then fill with batter. Bake for 20 minutes, or until light brown on top. Remove the muffins from the pan and let cool.

Makes 12 muffins

CHINESE WORM SKILLET DISH

1 pound ground beef or turkey
$^1/_2$ onion, chopped
3 or 4 stalks of celery, chopped
$^1/_2$ cup milk
1 (8-ounce) package frozen Chinese pea pods,
cooked and drained
1 can (10$^1/_2$ ounces) cream of chicken soup
$^1/_2$ cup boiled earthworms
1 can (5 ounces) crisp chow mein noodles

Add the ground meat, onion, and celery to a skillet and cook until the meat is browned and the vegetables are crisp-tender. Add the milk, pea pods, soup, and worms to the skillet. Mix and heat through. To serve, place some of the chow mein noodles on each plate. Top the noodles with some of the meat mixture.

Serves 4

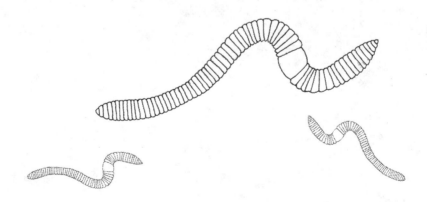

EARTHWORM SCRAMBLE

6 eggs

$1/3$ cup milk

Salt and pepper to taste

$1/4$ teaspoon garlic powder

1 tablespoon butter or margarine

$1/4$ cup chopped onion

$1/4$ cup green bell pepper

$1/3$ cup chopped tomato

$1/2$ cup sliced mushrooms

$1/4$ cup boiled, chopped earthworms

$1/3$ cup grated cheddar or American cheese

Beat together the eggs, milk, salt, pepper, and garlic powder. Set aside. Place the butter in a sauté pan and melt over medium-high heat. Add the onion and bell pepper to the melted butter. Cook until tender. Add the egg mixture, tomatoes, mushrooms, and worms. Cook until the eggs set. Serve on a plate topped with the grated cheese.

Serves 2 to 3

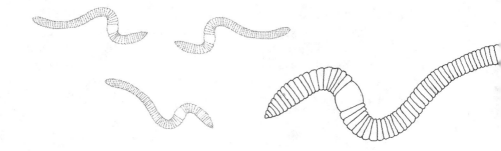

EARTHWORM MEATLOAF

1 1/2 pounds ground beef
1/2 cup boiled earthworms, finely chopped
1 envelope dry onion soup mix
1/2 cup evaporated milk
1/2 bell pepper, chopped
1 slice fresh bread, torn into bits

Mix all ingredients together and place in a loaf pan. Bake for 1 hour at 400°F. Enjoy.

Serves 4 to 6

CARAMEL EARTHWORM BROWNIES

1 package of your favorite brownie mix
(or your own homemade recipe)
2 tablespoons earthworm flour
1 cup chopped nuts
1/4 cup bottled caramel sauce

Combine the brownie mix with the earthworm flour and prepare according to package directions. Stir in the nuts.
Pour one-half to two-thirds of the batter into the pan. Drizzle the caramel sauce on top of the batter. Pour the remaining batter on top of the caramel sauce. Bake according to package directions.

Serves 8 to 10

Buying Guide
to Worms, Worm Products, and Organic Products

Beaver River
P.O. Box 94
W. Kingston, RI 02892
Phone/Fax: (401) 782-8747
www.merchantsbay.com/
beaverriver
*Worms, coir fiber, insert kits,
bins, and classroom materials*

C. K. Ventures Ltd.
P.O. Box 2052
Ladysmith, BC VOR 2EO
Canada
Metal worm bins

Flowerfield Enterprises
10332 Shaver Road
Kalamazoo, MI 49024
Phone: (616) 327-0108
*Worm bins, worms, worm videos,
and educational materials*

Grangettos Farm and
Garden Supply
1105 W. Mission Avenue
Escondido, CA 92025
Phone: (760) 745-4671
*Worms and organic products,
free catalog*

Organic Gardeners Resource and
Design Centre
315 S. Coast Hwy 101, Suite U32
Encinitas, CA 92024
Phone: (888) 514-4004
Email: curly@mill.net
*High quality organic fertilizers,
very large selection of organic
products, beneficial insects,
nontoxic pest controls, and
information for individuals
interested in environmentally
friendly gardening practices*

Solana Recyclers, Inc.
137 N. El Camino Real
Encinitas, CA 92024
Phone: (760) 436-7986
Fax: (760) 436-8263
Email: solana@adnc.com
*Worm bins, worms, compost
bins, literature, and technical
help. They also coordinate
Master Composter training
programs and composting
workshops.*

Simpson's Nursery
13925 Highway 94
Jamal, CA 91935
Worms and organic products

VermiCo
P.O. Box 1134
Merlin, OR 97532
Phone: (541) 476-9626
Fax: (541) 476-4555
Email: vermico@cdsnet.net
web site: HTTP://www.nwim.com/
vermico/
*Can-O-Worms, worms, and
vermiculture technology*

Walter Andersen's Nursery
3642 Enterprise
San Diego, CA 92110-3212
Phone: (619) 224-8271
Worms and organic products

Williams Worm Farm
Cyclone Manufacturing
14893 El Monte Road
Lakeside, CA 92040
(619) 443-1698
Worms, worm bins, and supplies

Worm World, Inc.
12425 NW CR 231
Gainesville, FL 32609
Phone: (352) 485-1235
Fax: (352) 336-3680
*Commercial worm bins, worms,
and worm products*

Yelm Earthworm and
Castings Farm
14741 Lawrence Lake Rd. S.E.
Yelm, WA 98597
Phone: (360) 894-0707
email: yelmworms@aol.com
*Worms, castings, and vermicul-
ture technology and support*

GUIDE TO ADDITIONAL INFORMATION

Newsletters

Casting Call

VermiCo's bimonthly newsletter. The primary focus of this newsletter is vermiculture, composting, soil fertility, and related issues of organic waste. Subscription rate is $18 for six issues.

Contact: VermiCo

    P.O. Box 1134

    Merlin, OR 97532

    Phone: (541) 476-9626

    Fax: (541) 476-4555

    Email: vermico@cdsnet.net

    Web site: http://www.nwim.com/vermico/

Worm Digest

Order a subscription to *Worm Digest*, a quarterly newsletter that bears the subtitle, "Worms Deepening Our Connection to Food and Soil." Articles deal with the promotion of sustainable agriculture through worms.

United States: $12/year (4 issues)

Back Issues: $3.50 each

Contact: Worm Digest

    Box 544

    Eugene, OR 97440-0544

    http://www.wormdigest.org/html

Some Informational Internet Sites

Vermicomposting Forum
http://www.oldgrowth.org/compost/wwwboard/
vermi/index.html

The Burrow
http://gnv.fdt.net/~windle/

BuyerTec Worm Growers Hints
http://www.ozemail.com.au/~rhee/worms.htm

The Earthworms as Fertilizer Factory
http://www.rain.org/~sals/worms.html

The Lowly Earthworm: The Gardener's Friend
http://www.ag.usask.ca/cofa/departments/hort/
hortinfo/misc/earthwor.html

The City Naturist—Earthworms
http://www.nysite.com/nature/fauna/earthworm.htm

Organic Farming
http://www.wdc.net/~smd/agorg.htm

GLOSSARY

ACIDIC—having characteristics of acids, such as a sour taste and testing lower than seven on a pH meter.

AERATION—getting oxygen into something (a worm bin) by mixing or turning.

AEROBIC—anything that requires oxygen. Aerobic bacteria need oxygen to live.

AGGREGATION—the clumping together of soil particles, which aids in aeration and water penetration.

ALBUMIN—a protein component for cocoons that feeds developing worms.

ALIMENTARY CANAL—the digestive tract of a worm.

ALKALINE—having the characteristics of alkali and testing higher than seven on a pH meter.

ANAEROBIC—anything that does not require oxygen. Anaerobic bacteria live in conditions that do not have oxygen.

ANECIC WORMS—worms that burrow deep into the soil but forage at night on the surface for freshly decaying plant or animal residues (example: nightcrawlers).

ANNELID—a worm belonging to the phylum Annelida; it has a segmented body and a coelom.

ANTERIOR—towards the front of a worm.

ANUS—the exit of the digestive tract, usually located near the rear of an animal.

ARTHROPOD—an invertebrate animal that is a member of the phylum Arthropoda, including insects, arachnids, and crustaceans.

BED-RUN—a term used to describe both adult and immature worms that you would buy together from a worm grower.

BEDDING—materials like newspaper and leaves used as an organic medium for worm composting.

BENEFICIAL—a term used to describe a plant or animal that contributes to the well-being of people or nature.

BREEDERS—a term referring to adult worms that have the ability to reproduce.

CASTINGS—worm manure or excrement.

CELLULOSE—a compound made of carbon, hydrogen, and oxygen that is the principal component of cell walls of plants. It is a complex carbohydrate.

CLITELLUM—a swollen region near the front end of an adult worm (such as a leech or earthworm) that produces mucus and makes the cocoon for the eggs.

COCOON—an egg case; animals such as moths and worms produce cocoons. Worm cocoons can carry from two to twenty worms each.

COELOM—a body cavity, found in annelids and other advanced animals, that is lined on all sides by a thin layer of cells and is a fluid-filled space between the earthworm's body wall and digestive system.

COMPOST—the end product or humus that is produced by the decomposer organisms in the composting process.

COMPOSTABLE MATERIALS—organic materials such as leaves and grass that will break down in a compost bin.

COMPOSTING—the process that occurs when organic matter is broken down by decomposer organisms into a nutrient-rich soil or humus.

CONSUMERS—organisms that obtain energy by eating other organisms or food particles.

CROP—a thin-walled sac just behind the earthworm's esophagus where food is stored briefly on its way through the digestive system.

CUTICLE—an outer body layer that is secreted by skin cells and helps protect the body from the environment.

DECOMPOSERS—organisms that digest and break down into simpler compounds the organic matter in the dead bodies of other plants and animals.

DECOMPOSITION—the process of breaking down organic matter into its basic compounds and elements, including nutrients needed for plant growth.

EARTHWORM—a segmented worm that belongs to the phylum Annelida.

ECOSYSTEM—an interacting dependent system that includes all plants, animals, and inorganic matter.

ENDEMIC—something that belongs to a particular locality or region.

ENDOGEIC WORMS—soil-dwelling worms that ingest soil and extract nutrition from degraded organic matter.

EPIDERMIS—the outer layer of an animal's body.

EPIGEIC WORMS—surface-dwelling worms that ingest freshly decaying plant or animal residues.

ESOPHAGUS—in worms, this is the food canal that leads from the mouth to the crop and gizzard.

FEED STOCK—a term used to describe food that is fed to worms or livestock.

FERTILIZER—a substance that is added to the soil to supply one or more plant nutrients; it can be either natural or man-made.

FOOD CHAIN—transfer of food energy through producers, consumers, and decomposers.

FOOD WASTES—food scraps; generally refers to uncooked fruit and vegetable scraps.

FOOD WEB—a complex, interlocking series of food chains.

GIZZARD—sac behind the earthworm's crop that has heavy muscular walls that grind up food with the help of grit swallowed by the worm.

HABITAT—a place where an organism lives.

HEAP—an unenclosed compost pile.

HERMAPHRODITE—an organism that has both male and female reproductive organs.

HIBERNATION—a state of dormancy, especially in winter.

HUMUS—finished compost, which has undergone a high degree of decomposition through the breakdown of plant and animal matter. It is stable, dark in color, and has a high water absorption and swelling capability.

HYDRATED LIME—also called calcium hydroxide, it is not recommended for use in a worm bin.

INFILTRATION—the ability of water to penetrate into soils.

INVERTEBRATES—organisms like insects and worms that do not possess a backbone.

LEACHATE—liquid that results from solid waste decomposition and which has extracted, dissolved, or suspended materials in it.

LEAF LITTER—the uppermost organic materials that are partly or not at all decomposed, on the surface of the soil.

LEAF MOLD—decomposed or mostly decomposed leaves.

MACROORGANISMS—organisms that can be seen without magnification.

MICROORGANISMS—organisms that are extremely small and cannot be seen without magnification.

MUCUS—a slimy substance produced by body cells.

MULCH—a layer of plant residues, such as partially decomposed plant materials, which is placed on top of garden beds and around plants and shrubs to hold in moisture.

NEMATODE—a roundworm belonging to the phylum Nematoda; these worms have a straight digestive tract that lies loose within a fluid-filled body space.

NOCTURNAL—an organism that is active during the night.

OLIGOCHAETE—an annelid, such as an earthworm, that has only a few bristles (setae) and has a clitellum.

ORDER—a subdivision of a class of organisms, comprising a group of related families.

ORGANIC MATTER—any organic material, derived from plants or animals, that is in a more or less advanced stage of decomposition.

OVERLOAD—to put more food into a worm bin than can be processed aerobically.

pH—a chemical term meaning "potential hydrogen." A pH scale measures the acidity to alkalinity of something on a scale from one to fourteen. Seven on the scale is neutral.

PARASITE—an organism that lives on or in the body of another living organism (host) for at least part of its life.

PARTHENOGENESIS—the production of offspring without fertilization.

PATHOGEN—an organism or virus that causes disease.

PEST—an organism that is harmful or annoying to humans.

PHARYNX—a muscular part of the digestive tract behind the mouth that is used to pump food into the digestive system.

PHEROMONE—a substance produced by one organism that influences the behavior or physiology of another organism of the same species.

PHOTOPHOBIC—an organism that hates light.

PHYLUM (PL. PHYLA)—a group of living things that share a common body plan.

PRODUCERS—organisms and plants that change energy from the sun into food energy, which can be used by consumers.

PROSTOMIUM—a fleshy pad that exists above the mouth of a worm. It aids in taking in food.

RECYCLE—to use something again and again.

REDWORM—a variety of earthworm suitable for vermicomposting, such as *Eisenia fetida*.

REGENERATE—to regrow a new part of the body that has been lost or hurt.

RODENT RESISTANT—anything designed or modified in such a way as to deter rodents from making a home in it.

SCIENTIFIC NAME—an internationally recognized Latin name for a species.

SCREENING—to sift out uncomposted matter and worms from the compost and castings.

SEGMENT—a section or part.

SEPTA—sheets of tissue that separate the bodies of annelid worms into separate segments.

SETAE—the spines or bristles of oligochaetes.

SOIL—a mixture of weathered rocks, sand, silt, clay, and organic material that covers the earth in a thin layer in which plants grow.

SOIL CONDITIONER—something that enriches the physical condition of soil and increases its organic content.

SPECIES—a group of individuals that are similar in structure and physiology and are capable of reproducing fertile off-spring.

VERMICAST—another word for castings produced from worms that have eaten and re-eaten the same material many times.

VERMICOMPOST—the end product from composting with worms. Vermicompost contains worm castings, broken-down organic matter, bedding, worm cocoons, worms, and other organisms.

VERMICOMPOSTER—a worm bin or a person who composts with worms.

VERMICULTURE—worm farming or raising earthworms.

WORM BIN—a container especially prepared for worms to live in and eat organic material.

BIBLIOGRAPHY

BOOKS

Appelhof, Mary. *Worms Eat My Garbage*, 2nd ed. Kalamazoo, MI: Flowerfield Press, 1997.

Ball, Jeff. *Rodale's Garden Problem Solver*, Emmaus, PA: Rodale Press, 1988.

Barnes, Robert D. *Invertebrate Zoology*, 2nd ed. Philadelphia, PA: Saunders Company, 1968.

Bogdanov, Peter. *Commercial Vermiculture: How To Build a Thriving Business in Redworms*, Merlin, OR: VermiCo., 1996.

Edwards, C. A. and J. R. Lofty. *Biology of Earthworms*, 2nd ed. New York: Halsted Press, 1977.

Keeton, William T. *Biological Science*, 2nd ed. New York: Norton & Company, Inc., 1972.

Martin, Deborah L. and Grace Gershuny, editors. *The Rodale Book of Composting*, Emmaus, PA: Rodale Press, 1992.

McGrath, Mike. *The Best of Organic Gardening*. Emmaus, PA: Rodale Press, 1996.

McLaughlin, Molly. *Earthworms, Dirt, and Rotten Leaves*. New York: Macmillan Publishing Company, 1986.

Nancarrow, Loren and Janet Hogan Taylor. *Dead Snails Leave No Trails*. Berkeley, CA: Ten Speed Press, 1996.

Patent, Dorothy Hinshaw. *The World of Worms*. New York: Holiday House, 1978.

Simon, Seymour. *Discovering What Earthworms Do*. New York: McGraw-Hill Book Company, 1969.

Taylor, Ronald L. and Barbara J. Carter. *Entertaining with Insects*, Yorba Linda, CA: Salutek Publishing Company, 1992.

Articles

Bambara, S. B. (Entomology) and R. L. Sherman. "Controlling Mite Pests in Earthworm Beds." Dept. of Biological & Agricultural Engineering, North Carolina State University, 1997.

Department of Animal Hygiene Paper Summary. "Effect of Organophosphorous Insecticides on Cholinesterase activity in Earthworms from Farmyard." Croatia: University of Zagreb, 1995.

Lyon, W. F. "Earthworm Enemies." HYG-2134-96. Ohio Cooperative Extension Service. 1996.

Martin, J. P, J. H. Black, and R. M. Hawthorne. "Earthworm Biology and Production." Leaflet 2828. University of California Cooperative Extension Service, July 1976.

Mathies, J. B. "Growing Earthworms for Fun and Profit," Zoology Series 1. Raleigh, NC: North Carolina Cooperative Extension Service, April 1974.

Townsend, Lee and Dan Potter (Entomology) and A. J. Powell (Agronomy), "Earthworms: Thatch-Busters." University of Kentucky, 1997.

Trotter, Donald W. "Earthworm Update." Sustainable Agriculture, Davis, CA: University of California, Summer 1994.

UC Cooperative Extension. "Harvesting Your Worm Compost." San Andreas, CA: 1996

World Resource Foundation. Information Sheets: "Preserving Resources through Integrated Sustainable Management of Waste," England: April 1997.

Internet Reference Sites

The Burrow at
http://gnv.fdt.net/~windle/

Buyer Tec Worm Growers Hints at
http:www.ozemail.com.au/~rhee/worms.htm

Compost Resource Page at http://www.oldgrowth.org/
compost/wwwboard/verigroup/index.html

Earthworms at
http://www.cog.brown.edu/gardening/worms.html

New Jersey Online's Yucky Site: Worm World at
http://dev01.nj.com/yucky/worm

The City Naturist-Earthworms at
http://www.nysite.com/nature/fauna/earthworm.htm

The Earthworms as Fertilizer Factory at
http://www.rain.org/~sals/worms.html

The Lowly Earthworm: The Gardener's Friend at
http://www.ag.usask.ca/cofa/departments/hort/
hortinfor/misc/ earthwor.html

Organic Farming at http://www.wdc.net/~smd/agorg.htm

Vermicomposters Forum at http://www.oldgrowth.org/
compost/wwwboard/vermi/data/2296.html

Worm Songs at
http://www.sci.mus.mn.us/slm/tf/w/worms/worms/
song.html

INDEX

Printed in the United States
by Baker & Taylor Publisher Services